Introduction to Industrial Motor Control

Introduction to Industrial Motor Control

Jay F. Hooper

Carolina Academic Press
Durham, North Carolina

Copyright © 2010
Jay F. Hooper
All Rights Reserved

On demand edition

ISBN-13: 978-1724524508
ISBN-10: 172452450X

Originally published by:

Carolina Academic Press
700 Kent St.
Durham, NC 27701
Telephone (919) 489-7486
Fax (919) 493-5668
www.cap-press.com

Printed in the United States of America

Contents

Foreword xi

Chapter 1 · Tools 3
   Screwdrivers 3
   Pliers 6
   Hammers 7
   Wrenches 8
   Cutters and Knives 8
   Strippers and Crimpers 9
   Levels 10
   Exercise 1.1 Stripping Wire 13
   Exercise 1.2 Crimping Wire 16

Chapter 2 · Safety and ON-OFF Circuits 23
   Fire 23
   LOTO 26
   Some Definitions 28
   Circuit Is Shut Off Twice 30
   Lock Out 30
   Tag Out 30
   Lock Out, Tag Out 30
   Real World Circuits Should Be Shut Off
     Four Times 31
   PPE 32
   Electrical Symbols 32
   Symbols in the Electrical World 33
     Switches 33
     Contacts and Switches with Flags 35
     Coils and Loads 39
     Overcurrent Protection and Disconnects 41
     Miscellaneous 42
   Symbols in the Electronic World 44
     Solid State Diode 44
     SCR 44
     Triac 44

Contents

LED	44
NPN Transistor	44
PNP Transistor	45
Regular Resistor	45
Pot	45
Photo-Eye	45
Chassis Ground	45
Earth Ground	45
Conductors	46
Double Meanings	46
L1 L2 L3 (power diagrams and schematics)	46
L1 L2 (control diagrams and schematics)	47
T1 T2 T3 (power diagrams and schematics)	47
Phases	48
Single Phase	48
Exercise 2.1	50
Chapter 3 · Electrical Troubleshooting and Manual Motor Starter Circuits	**53**
Percentages	53
A Satisfied Customer	54
Low Cost	54
Quick Turnaround	55
1-800-Number Troubleshooting	55
(1) Is It On or Off?	55
(2) Fuses and Circuit Breakers	56
(3) Visual Inspection and/or Smell	56
(4) Bad Terminations	57
(5) Ohm Meter Work	57
(6) Volt Meter and Ammeter Work	57
(7) The Tough Stuff	58
Exercise 3.1	59
Chapter 4 · Overcurrent Protection and Automatic Control Circuits	**61**
Current and People (Touching Hazard)	61
15 ma	62
150 ma	62
1.5 amps	62
15 amps	63

Current and Devices (Flash and Explosion
    Hazard)   63
    .5 amps   63
    5 amps   63
    50 amps   64
    500 amps   64
    5,000 amps   64
    50,000 amps   64
Residential and Office Space Current Safety
    Considerations   65
Industrial Current Safety Considerations   65
Industrial Overcurrent Protection and Time   66
Automatic Control Circuits   67
Exercise 4.1   68

Chapter 5 · Ohm Meter Testing   69
Continuity   69
Solenoid Test   70
    Open   70
    Short   70
    Value   71
    Leakage   71
Test for Switches and Relay Contacts   72
Practice Makes Perfect   72
Exercise 5.1   74
Mechanical Operation   74
Modified Solenoid Test   74
Open   75
Short   75
Value   76
Leakage   76

**Chapter 6 · Start-Stop Pushbutton Circuits
    (Latch-Unlatch)**   79
Three Wire Control Circuit   79
No and Low Voltage Protection Circuit   80
Latch-Unlatch (Latching)   80
N.O. vs. Open and N.C. vs. Closed   84
Exercise 6.1   86
Reed Relays   86
Control Relays   86

Contents

Power Relays	87
Contactors	87
Motor Starters	87

Chapter 7 · Solid State Switches — 89

Proximity Detectors	89
Photoeyes	91
Retro Reflective	92
Diffuse	92
Thru Beam	93
Convergent	94
Light Dark Operate	94
Multiple Hookup Details of Solid State Switching	94
Exercise 7.1	96
Exercise 7.2	97
Exercise 7.3	98

Chapter 8 · Wye and Delta — 99

Three Phase Transformers	99
Wye	99
Delta	101
Three Phase Motors	103
Single Phase Transformers	107
Single Phase Motors	108
K Factor	109
Exercise 8.1	110
Exercise 8.2	111

Chapter 9 · Time Delays — 113

On Delay	113
Off Delay	114
One Shot	116
Other Time Delays	117
PLC Timers	118
Exercise 9.1	119
Exercise 9.2	120
Exercise 9.3	121

Chapter 10 · Three Phase Reversing and Interlocking Starters	123
Mechanical Interlocking	123
Electrical Interlocking	124
PLC Interlocking	125
Power Circuit	126
Overloads	126
Fixed Mechanical and Thermal Overload Protection	127
Adjustable Mechanical and Thermal Overload Protection	127
Adjustable Solid State Overload Protection	128
Exercise 10.1	129
Exercise 10.2	130
Exercise 10.3	131
Chapter 11 · Variable Frequency Drives (VFDs)	133
VFD	133
L1, L2, and L3	134
Converter (AC to DC section)	135
DC Filter (DC bus section)	135
Inverter (DC to AC section)	135
T1, T2, and T3	136
Low Voltage Control Strip	136
24 VAC Control (HVAC)	136
0–10 Volt DC Analog Control	137
0–5 Volt DC Analog Control	137
5 Volt DC Digital Control	137
4–20 ma. DC Analog Control	137
4–20 ma. DC Digital Control	137
Installation Issues (Small Terminals)	137
Installation Issues (Confusing Lingo)	138
Installation Issues (Whiskers)	138
Installation Issues (Wrong Voltage)	138
Installation Issues (Wrong Parameters)	139
Installation Issues (Bad Plant Input Power Circuit Voltage Levels)	140
Exercise 11.1	141
Exercise 11.2	143

Contents

Chapter 12 · Servos and Intermediate
                 Troubleshooting   145
   Resolvers and Synchros   145
   Encoders   146
   Stepper Motors   147
   Servo Motors   147
   Intermediate Troubleshooting   147
   Using Voltmeters   149
   Using Ammeters   150
   Using Meggers   151
   Using Rotation Meters   151
   Using Oscilloscopes   152
   In Conclusion   152

Appendix   155
   Selected Component Pictures — 2 Views   155

Foreword

The "we" in this book refers to my two good friends, Dick Pipan and Fred Gunter, and me. They were my assistants in running this seminar for industry after many years of teaching it myself in the N.C. community college system. The book focuses on the practical troubleshooting considerations of working on motor controls and three phase motors. The book is designed to get you up to speed quickly. You will be able to solve two-thirds or more of all the electrical problems that you will ever encounter in a very short time frame.

The book is heavily oriented to making you a great troubleshooter. In the actual working classroom or lab, four-fifths of the time is spent building the circuits in the exercises. The reason that I do that is not to teach about the individual circuit *per se*. I want to generate teachable moments. I want to have random glitches pop up all the time that classes can use as troubleshooting opportunities. Anything that you can possible think of that can go wrong with the components or the circuits will. To enhance this effect, I use a combination of old and new parts. I sort the parts before each new class, but I do not discard any parts or components that I suspect may be bad. I only mark or throw away student-identified defective parts.

To maintain additional control of safety in the lab setting I do not allow anyone to put electricity above 120 VAC on a circuit without first testing for dead shorts.

Jay F. Hooper
Salisbury, N.C.

Introduction to Industrial Motor Control

Tools 1

Before we begin the "hands-on circuit building" part of this book, it will be necessary to cover hand tool usage and safety. In this chapter, we will be covering hand tools. What we are going to do is to go through a few of the most common types of tools that an electrical technician or a maintenance mechanic would have in order to work on everyday industrial electrical control circuits and light duty electrical power circuits.

Screwdrivers

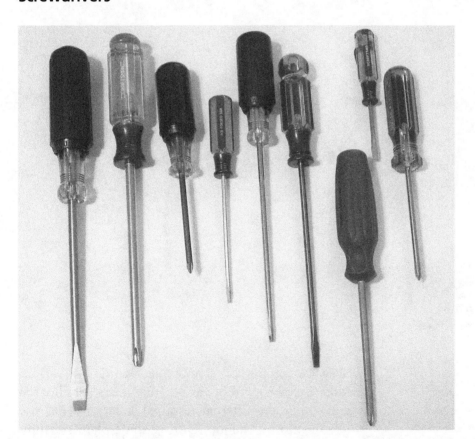

In front of you in your tool pouch you should see various types of screwdrivers of varying lengths and diameters. The two most common are the Phillips head screwdriver and the flat head or regular screwdriver.

**Illustration 1.1
various screwdrivers**

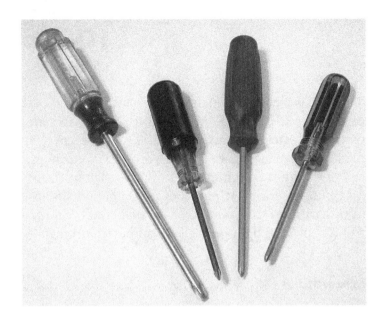

Illustration 1.2 Phillips head screwdrivers

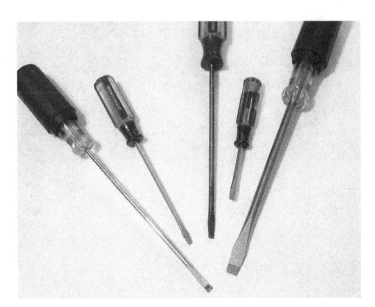

Illustration 1.3 regular screwdrivers

The two types of screwdrivers come in various widths (regular) and in various diameters (Phillips) in order to best fit different terminal situations that you will run into on the job. During some plant startups, we have worked on machines with European factory reps. They were so particular in using a screwdriver that not only did you have to pick out the proper screwdriver and tip for the job, but you had to dip your finger in oil. You then touched the threads of the screw for lubrication and rust prevention before driving the screw home.

In electrical work, as in a lot of other things in life, there are rules of thumb that people follow, as well as a right way and a wrong way of doing things. There is also a difference between the textbook way and the on-the-job way of doing things. We will expose you to the types of things that you may run into in a real job situation.

In the case of screwdrivers (pictured in illustration 1.4), most people call these screwdrivers "Phillips head screwdrivers" on the job. However, if you have to order the screwdrivers out of a catalog, the ones with a blunt end are Phillips head screwdrivers, and the ones with sharp ends are Reed and Prince screwdrivers.

The major issue that you run into with screwdrivers is that lots of people misuse them as chisels, levers, punches, or pry bars. In most trades, you need to be professional and use the tool for its originally designed and intended use. In the industrial world, if you are a mechanic, you should be using a screwdriver as a screwdriver. If you need a chisel, then go get a chisel. You do not need to use your screwdriver as a chisel. If you need a pry bar, then go get a pry bar. You don't need to use your giant screwdriver as a pry bar.

Illustration 1.4 screwdriver tips Phillips (left side) and Reed and Prince (right side)

If we walk into any industrial plant and even before we get out on the factory floor if we see maintenance mechanics walking around with a pair of pliers and a screwdriver in their back pocket we would know that your machines would be in terrible shape. Sight unseen. Your machines would be in absolutely terrible shape. If you are a mechanic, a pair of pliers is a tool of last resort. There are only two trades that usually use a screwdriver and Channel Lock™ pliers as a first choice tool, electricians and pipe fitters.

In general, it is completely acceptable for electricians to use a giant screwdriver and a hammer for putting on locknuts when they are running conduit, and for using Channel Lock™ pliers to deburr thin wall conduit and to tighten up nuts on the fittings of that same conduit.

In general, it is completely acceptable for pipe fitters to use a giant screwdriver for lining up flanges when they are running pipe, and for using a Channel Lock™ pliers to tighten up nuts on small fittings and to hold small diameter pipe. If you are a maintenance mechanic, you need to use a wrench, a chisel, and a line up punch to do these things. Having the proper wrench and the proper tools is the main "tool rule" for mechanics.

Pliers

The most common type of pliers that you will use in electrical work are the diagonal cutting pliers

**Illustration 1.5
various pliers**

(often called diagonal cutters or dykes), the long nose pliers (most often called needle nose pliers), the tongue and groove pliers (most often called Channel Locks™ after the company of the same name), lineman's pliers (also called side cutting pliers) and locking pliers (also called Vice Grips™ made by a company called Irwin).

The diagonal cutter is designed to cut soft copper or aluminum control wire, not steel. If you cut coat hangers or other steel wire with a diagonal cutter, you will ruin the blade surface and put pock marks and/or indentations into it. The long nose pliers are most often used to hold or retrieve very small objects or to hold control wire. The Channel Lock™ pliers are most often used to tighten nuts on electrical conduit. The lineman's pliers are most often used for manipulating or holding onto power wiring.

Hammers

The most common hammers in industrial electrical work are the ball peen hammer and the soft face hammer. The claw hammer is used primarily in residential electrical work.

**Illustration 1.6
various hammers**

1 · Tools

The ball peen hammer is used for striking objects and for setting small anchors in walls and floors. The soft face hammer is used to hit an enclosure cover or other piece of equipment when you do not want to mar or dent the surface of that item too much.

Wrenches

The most common type of wrench in electrical control wiring is the adjustable wrench. This is a wrench that was designed for low torque (light duty turning and holding) applications. It is not for heavy duty work. Heavy duty turning is where the adjustable wrench gets its other floor name, the knuckle buster.

Illustration 1.7
adjustable wrench

Cutters and Knives

Some cutters are also called pliers in electrical work. The most common types of cutters in industry are the diagonal cutter (also called diagonal cutting pliers), the box cutter, the electricians knife (also called a pocket knife), the side cutter (also called lineman's pliers), a combination cutter (also called a wire stripper, a machine bolt cutter, or a wire cutter combo), and a cable cutter. We know that this is confusing, but remember we are just reporting on what it is that you are going to run into, in the field. If someone says to hand them a dyke, then you will need to know what he or she is talking about.

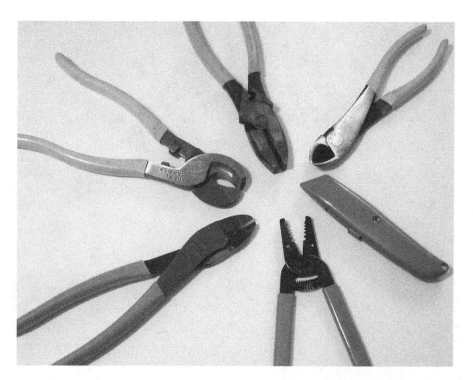

Illustration 1.8 various cutters and a knife

The diagonal cutter is used to cut copper wire. The box cutter and the pocket knife are used to cut insulation, usually splitting and then cutting it. The side cutter is used to cut power wiring (generally greater then 14 gauge solid or 18 gauge stranded). The cable cutter is generally used to cut copper and aluminum wire and cable bigger then 12 gauge.

Strippers and Crimpers

Strippers and crimpers usually come in a combination with other tools at the control wiring level (up to 14 gauge). More will come on this topic in the two exercises that follow at the end of this chapter. The most common type of combination is the crimper, cutter, bolt cutter, wire stripper combination tool (also known as a stripper), the wire stripper and cutter combination tool (also known as a stripper), and the combination cutter and crimper tool (also known as a stripper or as a diagonal cutter or as a side cutter).

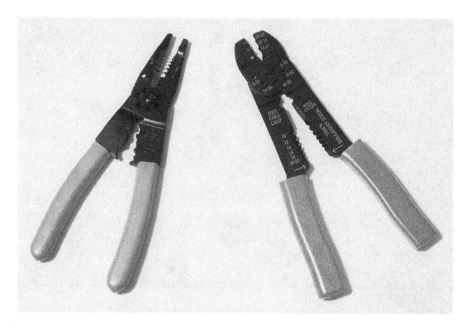

**Illustration 1.9
stripper/crimper combos**

Levels

In electrical work, it behooves you to have a 6 inch level (also known as a torpedo level) in your bag of tools. This is a very useful item if you are helping to mount electrical control boxes or outlet boxes. You do not want to end up with a wall of boxes that look like this:

**Illustration 1.10
uneven electrical boxes
mounted on a wall**

If your job looks this bad, then you or your company will not be getting any new installation work anytime soon. Last but not least, fuse pullers. We cannot begin to tell you how many times we have seen people pulling out fuses with a screwdriver, with a pair of pliers, or even with their fingers. It takes about 40,000 volts of electricity to jump about 1 cm. (about half an inch) in air. At less then $3 each, a pair of fuse pullers can easily keep your fingers many inches away from any potential electrical shock.

**Illustration 1.11
fuse pullers**

There are many stories floating around about how someone working close to live equipment or a live transformer died because the voltage "just jumped out" and hit them. Other than lightning and some experimental ASW military gear, there are not a whole lot of situations where electricity just jumps out and gets someone. Generally, someone's hand or other body part wanders around and the person inadvertently touches the high voltage equipment while they are working. Then, as they are falling backward, they pull an arc with them through a plasma tube (i.e. it jumped out and got them).

This is similar to what happens in a Jacob's ladder electrical display. You start out with a 20K to 40K volt neon tube transformer hooked up to two electrodes placed about a ¼ inch or so apart at the bottom, and about four to five inches apart on top. When the power is turned on, the arc strikes at the ¼ inch bottom air gap. A plasma tube forms. The arc heats it up. The plasma tube, being full of hot air, rises. The arc eventually reaches a length of 5 or more inches at the top. The arc goes out when it runs out of electrode. The process starts over. *Electricity just doesn't jump out and get you.* You touch it by mistake.

Again, this is not a comprehensive list of all of the tools that you will use, just an overview of the most common ones used in industrial control wiring.

Exercise 1.1
Stripping Wire

Unravel a piece of size #14 gauge stranded control wire THHN insulation or equivalent from a roll of wire. Cut two pieces of wire each about 12 inches long. Strip a ¼ inch of insulation from the end of one of the wires using a combo stripper that looks like this:

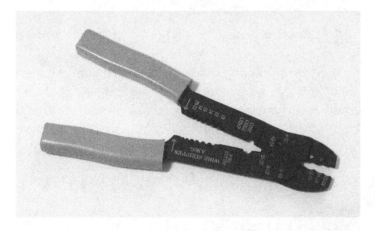

**Illustration 1.12
the drug store combo stripper**

Strip a ¼ inch of insulation from the end of the other wire using a combo stripper that looks like this:

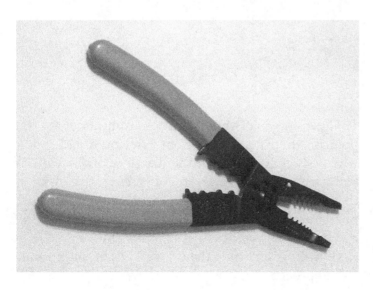

**Illustration 1.13
the electrical supply house combo stripper**

Exercise 1.1

Which stripper is easier to use? Remember you can not end up with any ringing of the wire. It's the electrical supply house one, wouldn't you say?

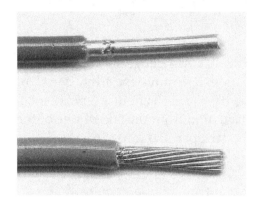

Illustration 1.14 ringing in stripped stranded and stripped solid copper wire

This wire ringing situation is a problem because ringing causes a resistance in the wire at that spot and could easily cause premature failure of your electrical circuit at that point in the wire. You also cannot have any missing strands in your wire.

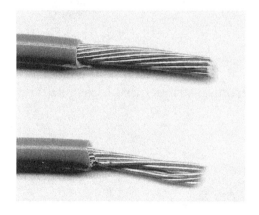

Illustration 1.15 missing strands in #14 stranded copper control wire (bottom wire)

These missing strands cause overheating at that point in the wire. The current carrying capacity of a wire is directly related to the diameter of the wire. If you have missing strands, the diameter of the wire at that point is less than normal, which could easily cause premature failure of your electrical circuit because of overheating at that point.

How do you correct these problems? It takes a four step approach:

(1) Choose the right tool. Use the electrical supply house stripper combo. Choosing to use the drug store stripper combo tool can allow you to make an excellent crimp, but it is a sorry excuse for a stripper. The rule of thumb for combo tools is that if the part of the tool that you want to use is in front of the pivot point of the tool, then you have a green light. If the part of the tool that you want to use is in back of the pivot point of the tool, then you have a red light.

(2) Choose the right hole. If you are after rabbits, look for rabbit holes. Do not go into a hole large enough to be the entrance to a bear's den. If you are going to strip #14 wires then use the #14 hole on the tool or the #12 hole depending on the type of insulation on the wire that you have in your hand.

(3) Choose the right angle. You must line up your wire at the correct angle to the plane of the tool.

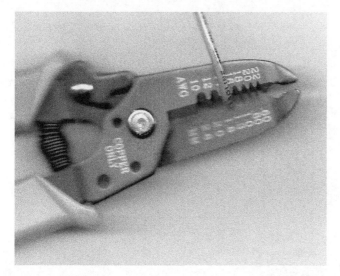

Illustration 1.16 strip angle

(Is this the right angle? No. This person is pulling on the wire up towards the top of the picture far too much. It needs to be pulled out of the plane of the book towards your chest.)

(4) Pull on the wire evenly. After closing the tool around the wire, pull evenly at the correct angle to strip the wire.

Exercise 1.2
Crimping Wire

If you are using solid wire and you wish to put a small diameter wire under a screw on a terminal strip, then you can use your long nose pliers. Make a button hook in the wire and put it under a terminal strip screw.

Illustration 1.17
a solid wire button hook

If you are using very small diameter stranded wire (like on a sensor of some sort), then you can tin the end of the wire with solder. Make a button hook and then hook your wire up under a terminal strip screw.

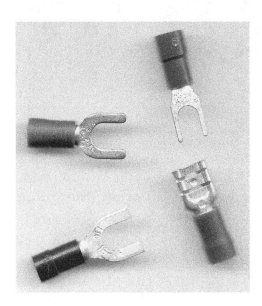

Illustration 1.18
#14 spade lug connector overview

If you are using larger diameter stranded control wiring (around 20–14 gauge) and smaller diameter power wiring (around 14–10 gauge), then you generally use connectors (also called crimps).

Take one of the stripped wires from exercise 1.1 and cut off both of the ends. Get a spade connector for #14 wire. Look closely at the spade connector. You should see a metal fork at one end and a plastic barrel at the other end.

Take a closer look at the spade lug connector. You should see four points starting from the left and going to the right:

(1) The beginning of the plastic barrel insulation.
(2) The beginning of the spade connector metal barrel as evidenced by a bump on the surface of the plastic barrel that covers it.
(3) The end of the metal barrel and the plastic barrel.
(4) The end of the metal spade lug.

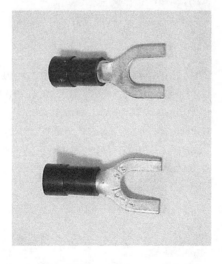

Illustration 1.19
#14 spade lug connector closeup view

Strip the wire off the end of the #14 control wire equal in size to the distance between points 2 and 3 (point 2 is the beginning of the metal part of the spade connector barrel, as evidenced by a bump on the surface of the plastic barrel that covers it, and point 3 is the end of the metal and the plastic barrel).

After stripping the wire, you can check if you have done it correctly by inserting the wire into the end of

Exercise 1.2

the plastic barrel until the insulation on the wire hits a shoulder on the inside of the barrel and stops the wire insertion. If the wire is even with point 3 or just pokes out a little then you are OK.

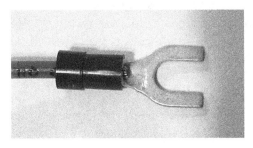

Illustration 1.20
OK wire strip

If the wire insertion stops on the inside shoulder but the bare part of the wire is not at point 3, then take the wire out, strip some more insulation off, and try again.

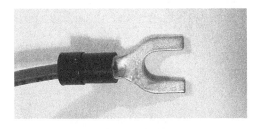

Illustration 1.21
redo the wire strip

If the wire inserts clear through the spade connector, then you need to try the next smaller size of connector or change to a different manufacturer's connector of the same size. There are often slight size variations in the manufacturing process in order to fit different insulation types and thicknesses.

Illustration 1.22
downsize the crimp

If the wire does not insert through to the internal shoulder of the spade connector, then you need to try the next larger size of connector or change to a different manufacturer's connector of the same size, as there are often size variations in the manufacturing process.

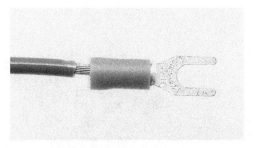

Illustration 1.23 upsize the crimp

You have now prepared the wire and the spade connector (the crimp). You now need to crimp the two together. In order to do this, choose the insulated crimping tool, also known as the drug store stripper combo tool, and then crimp the connection between points 2 and 3 on the spade connector. Your hand pressure is enough to mechanically make a gas tight connection on the inside of the spade connector. In some applications where the crimp will be stressed by a lot of vibration, you may be required to crimp twice.

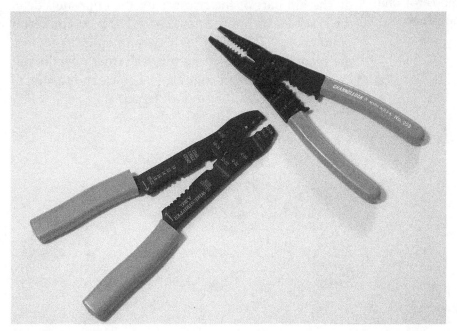

Illustration 1.24 insulated crimper combos

There is also a different type of crimper called the uninsulated crimper that is used a lot because it is easier on the hands when you are crimping. Over the years it has even been used on insulated connectors.

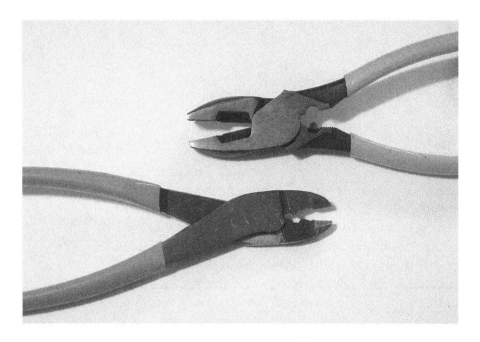

Illustration 1.25 crimper combos for uninsulated connectors

To crimp with this type of crimper, you need to line up the hill part of the crimper on the opposite side of the crack (seam) of the metal barrel of your connector (your crimp). Then, crimp between points 2 and 3 on the spade connector. Or, to put it another way, the valley of your crimper is to the metal crack and the hill of your crimper is to the smooth metal.

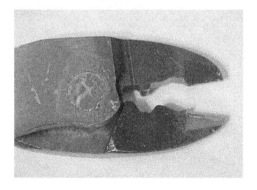

Illustration 1.26 hill and valley

You then apply pressure and get a good crimp. While this does somewhat crack the insulation, it has become acceptable for control wiring and light duty power wiring (under #10 gauge) in a lot of plants. The important thing is getting the gas-tight connection with the metal barrel and the wire.

Illustration 1.27
good but slightly ugly crimp

When you pick up the spade lug connector and look at it closely, it will be very easy to see where the crack or seam is in the metal barrel of your crimp. It is also relatively easy to see where the metal barrel begins and ends under the plastic barrel on the spade lug connector.

Safety and ON-OFF Circuits 2

We are going to cover some safety items first. If you have

Fire

1. First, report it or pull the alarm.
2. Get your people out.
3. If you then consider fighting it, you need to know what ABCDK is about.

Class A
A fire extinguisher labeled with **letter "A"** is for use on Class A fires. Class A fires involve ordinary combustible materials such as cloth, wood, paper, rubber, and many plastics.

Ordinary Combustibles

Class B
A fire extinguisher labeled with **letter "B"** is for use on Class B fires. Class B fires involve flammable and combustible liquids such as gasoline, alcohol, diesel oil, oil-based paints, lacquers, and flammable gasses.

Flammable Liquids

Class C
A fire extinguisher labeled with **letter "C"** is for use on Class C fires. Class C fires involve energized electrical equipment.

Electrical Equipment

Class D
A fire extinguisher labeled with **letter "D"** id for use on Class D fires. Class D fires involve combustible metals such as magnesium, titanium, and sodium.

Combustible Metals

Class K
A fire extinguisher labeled with **letter "K"** is for use on Class K fires. Class K fires involve vegetable oils, animal oils, or fats in cooking appliances. This is for commercial kitchens, including those found in restaurants, cafeterias, and caterers.

Combustible Cooking

Illustration 2.1
ABCDK

2 · Safety and ON-OFF Circuits

ABCDK are fire extinguisher codes. A is an extinguisher that is used for a fire that leaves an ash (wood, paper, clothing, etc.). B is an extinguisher that is used for flammable liquids (kerosene, gasoline, oil, etc.). C is an extinguisher that is used on electrical fires or on electrical equipment. D is an extinguisher that is used on combustible metals (aluminum, magnesium, etc.). K is used for heavy grease fires (around deep fat fryers cooking French fries, chicken, etc.).

The D extinguisher is not that well known. How do you put out a metal fire if your plant does not have a D extinguisher? You can put out a metal fire using dry sand (like the bags of play sand that they sell inside the main building of a large home improvement store). DO NOT use sand stored outdoors to try and put out a metal fire. That type of sand contains moisture in the form of H_2O or water that will cause the metal to splatter and the fire to scatter. The water could also disassociate into oxygen and hydrogen in the intense heat of the fire and then recombine, making the fire worse.

A lot of computer rooms use automated fire extinguisher equipment that can—and will—go off when a fire is detected. Since most computer and telecom equipment tends to be kept in small enclosed rooms, you have the additional hazard of possibly being asphyxiated by the extinguishers going off and displacing the oxygen in the room.

The most dangerous sign in fire situations that we find in any plant is this one:

NOT AN EXIT

Illustration 2.2
bad sign

The reason that this sign is so dangerous is that when people are under extreme stress, all that they see is the word "Exit." They do not see the other words on the sign.

Exit doors should never be chained or locked shut. Double doors to offices or eating areas should never have one half of the doorway locked to keep in or to keep out heat or cold, etc. If this occurs at your workplace, then in a bad smoky fire, people will be piling up at the door. They could break arms and legs, and possibly die at these half blocked doors.

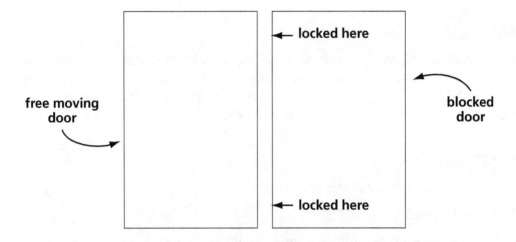

Illustration 2.3 blocked double door

You should also know where the nearest exit and fire extinguisher are when you are at a work area within your plant. You should know how to call 911 from your workplace if you need to. Remember anybody's cell phone will call out on 911, even if the landlines where you are located need you to dial a 9 or some other number first. The cliché says that "speed kills," but in fire situations, speed is of the essence. Get the fire department on the way first. It's the ultimate backup plan.

LOTO

In the OSHA standard 1910.147 on Lockout Tagout (LOTO), we see some general considerations:

Title: The Control of Hazardous Energy (lockout/tagout).

1910.147(a)

Scope, application and purpose—

1910.147(a)(1)

Scope

1910.147(a)(1)(i)

This standard covers the servicing and maintenance of machines and equipment in which the **unexpected** energization or start up of the machines or equipment, or release of stored energy could cause injury to employees. This standard establishes minimum performance requirements for the control of such hazardous energy.

1910.147(a)(1)(ii)

This standard does not cover the following:

1910.147(a)(1)(ii)(A)

Construction, agriculture and maritime employment;

1910.147(a)(1)(ii)(B)

Installations under the exclusive control of electric utilities for the purpose of power generation, transmission and distribution, including related equipment for communication or metering; and

1910.147(a)(1)(ii)(C)

Exposure to electrical hazards from work on, near, or with conductors or equipment in electric utilization installations, which is covered by Subpart S of this part; and

1910.147(a)(1)(ii)(D)

1910.147(a)(1)(ii)(D)

Oil and gas well drilling and servicing.

1910.147(a)(2)

Application.

1910.147(a)(2)(i)

This standard applies to the control of energy during servicing and/or maintenance of machines and equipment.

1910.147(a)(2)(ii)

Normal production operations are not covered by this standard (See Subpart O of this Part). Servicing and/or maintenance which takes place during normal production operations is covered by this standard only if:

1910.147(a)(2)(ii)(A)

An employee is required to remove or bypass a guard or other safety device; or

1910.147(a)(2)(ii)(B)

An employee is required to place any part of his or her body into an area on a machine or piece of equipment where work is actually performed upon the material being processed (point of operation) or where an associated danger zone exists during a machine operating cycle.

Note: **Exception to paragraph (a)(2)(ii):** Minor tool changes and adjustments, and other minor servicing activities, which take place during normal production operations, are not covered by this standard if they are routine, repetitive, and integral to the use of the equipment for production, provided that the work is performed using alternative measures which provide effective protection (See Subpart O of this Part).

1910.147(a)(2)(iii)

This standard does not apply to the following:

1910.147(a)(2)(iii)(A)

Work on cord and plug connected electric equipment for which exposure to the hazards of unexpected energization or start up of the equipment is controlled by the unplugging of the equipment from the energy source and by the plug being under the exclusive control of the employee performing the servicing or maintenance.

1910.147(a)(2)(iii)(B)

Hot tap operations involving transmission and distribution systems for substances such as gas, steam, water or petroleum products when they are performed on pressurized pipelines, provided that the employer demonstrates that—

1910.147(a)(2)(iii)(B)(1)

continuity of service is essential;

1910.147(a)(2)(iii)(B)(2)

shutdown of the system is impractical; and

1910.147(a)(2)(iii)(B)(3)

documented procedures are followed, and special equipment is used which will provide proven effective protection for employees.

1910.147(a)(3)

Purpose.

1910.147(a)(3)(i)

This section requires employers to establish a program and utilize procedures for affixing appropriate lockout devices or tagout devices to energy isolating devices, and to otherwise disable machines or equipment to prevent unexpected energization, start up or release of stored energy in order to prevent injury to employees.

1910.147(a)(3)(ii)

When other standards in this part require the use of lockout or tagout, they shall be used and supplemented by the procedural and training requirements of this section.

Some Definitions

Affected employee. An employee whose job requires him/her to operate or use a machine or equipment on which servicing or maintenance is being performed under lockout or tagout, or whose job requires him/her to work in an area in which such servicing or maintenance is being performed.

Authorized employee. A person who locks out or tags out machines or equipment in order to perform servicing or maintenance on that machine or equipment. An affected employee becomes an authorized employee when that employee's duties include performing servicing or maintenance covered under this section.

Capable of being locked out. An energy isolating device is capable of being locked out if it has a hasp or other means of attachment to which, or through which, a lock can be affixed, or it has a locking mechanism built into it. Other energy isolating devices are capable of being locked out, if lockout can be achieved without the need to dismantle, rebuild, or replace the energy isolating device or permanently alter its energy control capability.

Energized. Connected to an energy source or containing residual or stored energy.

Energy isolating device. A mechanical device that physically prevents the transmission or release of energy, including but not limited to the following: A manually operated electrical circuit breaker; a disconnect switch; a manually operated switch by which the conductors of a circuit can be disconnected from all ungrounded supply conductors, and, in addition, no pole can be operated independently; a line valve; a block; and any similar device used to block or isolate energy. Push buttons, selector switches and other control circuit type devices are not energy isolating devices.

Energy source. Any source of electrical, mechanical, hydraulic, pneumatic, chemical, thermal, or other energy.

Hot tap. A procedure used in the repair, maintenance and services activities which involves welding on a piece of equipment (pipelines, vessels or tanks) under pressure, in order to install connections or appurtenances. it is commonly used to replace or add sections of pipeline without the interruption of service for air, gas, water, steam, and petrochemical distribution systems.

Lockout. The placement of a lockout device on an energy isolating device, in accordance with an established procedure, ensuring that the energy isolating device and the equipment being controlled cannot be operated until the lockout device is removed.

Lockout device. A device that utilizes a positive means such as a lock, either key or combination type, to hold an energy isolating device in the safe position and prevent the energizing of a machine or equipment. Included are blank flanges and bolted slip blinds.

Normal production operations. The utilization of a machine or equipment to perform its intended production function.

Servicing and/or maintenance. Workplace activities such as constructing, installing, setting up, adjusting, inspecting, modifying, and maintaining and/or servicing machines or equipment. These activities include lubrication, cleaning or unjamming of machines or equipment and making adjustments or tool changes, where the employee may be exposed to the **unexpected** energization or startup of the equipment or release of hazardous energy.

Setting up. Any work performed to prepare a machine or equipment to perform its normal production operation.

Tagout. The placement of a tagout device on an energy isolating device, in accordance with an established procedure, to indicate that the energy isolating device and the equipment being controlled may not be operated until the tagout device is removed.

Tagout device. A prominent warning device, such as a tag and a means of attachment, which can be securely fastened to an energy isolating device in accordance with an established procedure, to indicate that the energy isolating device and the equipment being controlled may not be operated until the tagout device is removed.

And the rest of the story which we will summarize here:

In an industrial plant electrical situation, the minimum procedure that OSHA requires is simply to have a policy. It needs to be clear and employees need to follow it. Your company can have a lock out policy. Your company can have a tag out policy. Your company can have a lock out, tag out policy. Your particular LOTO policy usually depends on the type of industry that you are in.

Circuit Is Shut Off Twice

OSHA at a minimum requires that you ensure that the circuit is dead in two ways. One is to lock out, tag out, or lock out AND tag out the main disconnect for the machine or the circuit that you will be working on. Secondly, you have to verify that the circuit is dead.

Lock Out

If you use lock out, you need to be in control of the key that you use to lock out the disconnect.

Tag Out

If you use tag out, your name must be on the uniformly printed tag used by your company and the tag must be secured at the disconnect to the place that you would normally put a lock on. That secured tag must by held on by a band of material that can withstand at least 50 lbs. of force (such as a TyRap™).

Lock Out, Tag Out

If you use lock out, tag out, you need to be in control of the key that you use to lock out the disconnect. Your name must also be on the uniformly printed tag used by your company.

Real World Circuits Should Be Shut Off Four Times

In the general scheme of things, the chances that you will touch a live circuit and get shocked boil down to both your professionalism and mathematical probability. Let us predict your future. On average, one out of about every 100 people that you know will die in a car accident. One out of about every 325 people that you know will die from a gunshot wound. One out of about every 1100 people that you know will die in a fire or from smoke inhalation from that fire.

How do we know that? It's called data and statistics. There are people in the insurance industry that could tell you the number of people that live in North Carolina over the age of 50 that will get broken left thumbs next year. They just can't tell you who it is going to happen to.

This is our observation of how to put the odds in your favor in electrical work, no matter what trade group you may be working with.

We turn off the machine by locking it out, tagging it out, or shutting it off. We turn off the machine again by locking it out, tagging it out, or shutting it off at a second place. We verify that the machine is off. We then use a meter to test the circuit or short it out with a screwdriver before actually touching it with our hands. It is important to do these things *in this order*.

Some coworkers may turn the machine off at the stop button. Then they turn the machine off at the E-stop, then at the disconnect and then test the circuit with a meter before touching it. The most important thing to remember is to do something to shut the machine off at least four times. If you shut it off just once, guess what? You may have made a mistake and shut the wrong machine off.

Our observation in the electrical troubleshooting side of the trade is that on average, a one-step shut off procedure gives you a one in 10 chance of getting shocked. On average, a two-step shut off procedure gives you a one in 100 chance of getting shocked. A three-step shut off procedure gives you a one in 1,000 chance of getting shocked and a four-step shut off procedure gives you a one in 10,000 chance of getting shocked.

Anyone who works around electricity and tells you that they have never been shocked is lying to you. It is not a matter of when you are going to get shocked, it is a matter of how often and how severe. If you are professional, follow a LOTO procedure, and shut down the piece of equipment at multiple points, then your chances of getting seriously shocked or killed are very, very low. If you are a cowboy around electricity, then you will get hurt.

In working around electrical circuits, on average, one out of about every 5,000 people you know will die from an electrical shock.

PPE

Personal Protective Equipment (PPE) is governed by NFPA 70E. If you need to go inside an electrical cabinet where there are live electrical circuits, you must wear some sort of PPE to protect you against arc flash, arc blast, and other live electrical hazards. The easiest way around having to wear PPE is to shut off the electricity to the electrical cabinet that you have to go into. Then, there is no live electrical circuit inside. You will not need the PPE gear.

Electrical Symbols

In order to work on and troubleshoot electrical control and power circuits, you need to be able to read electrical prints and interpret symbols. The following is a list of typical NEMA (U.S.) symbols. You need to learn them. To compile this list, we went through some typical large electrical prints and electrical textbooks and pulled out some of the most common symbols. The list is not comprehensive, but it covers about 90 percent of what you would typically run into. We tell people to go look up the other symbols or ask someone you work with. IEC (European) symbols will be covered in the appendix in the back of this book or in the instructor's test bank.

Symbols in the Electrical World

Switches

Illustration 2.4
NEMA electrical symbols

Figure 1.

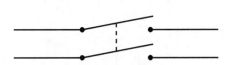

DPST
two position switch
(double pole
single throw)

3PDT
two position switch
(three pole
double throw)

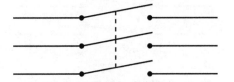

3PST
two position switch
(three pole
single throw)

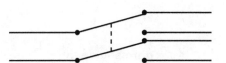

DPDT
two position switch
(double pole
double throw)

SPDT
two position switch
(single pole
double throw)

Figure 2.

SPST
two position switch
(single pole single throw)

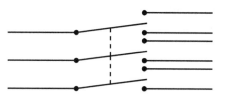

3PDT
three position center off switch
(three pole double throw)

normally open
momentary contact
single pole
pushbutton switch

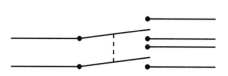

DPDT
three position center off switch
(double pole double throw)

SPST
normally closed switch
(two position)

Contacts and Switches with Flags

Figure 3.

normally closed
off delay
timed contact

normally open
on delay
timed contact

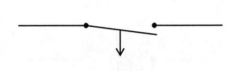

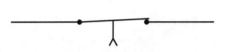

normally open
off delay
timed contact

normally closed
on delay
timed contact

Figure 4.

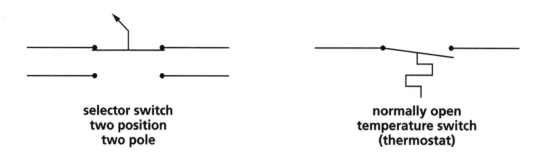

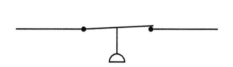

normally closed
pressure (or vacuum)
switch

selector switch
three position
two pole

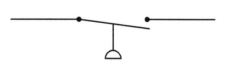

normally open
pressure (or vacuum)
switch

Figure 5.

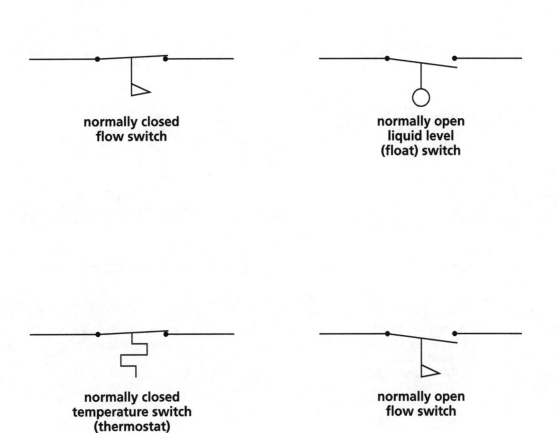

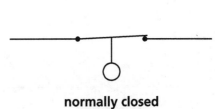

Figure 6.

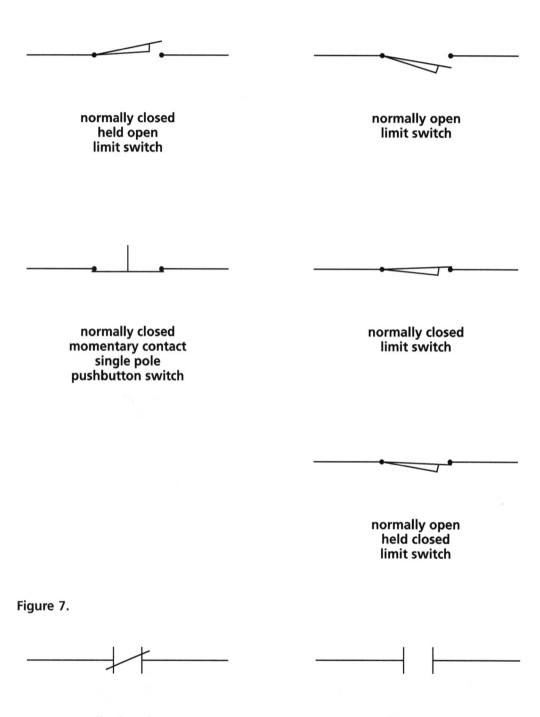

Figure 7.

Coils and Loads

Figure 8.

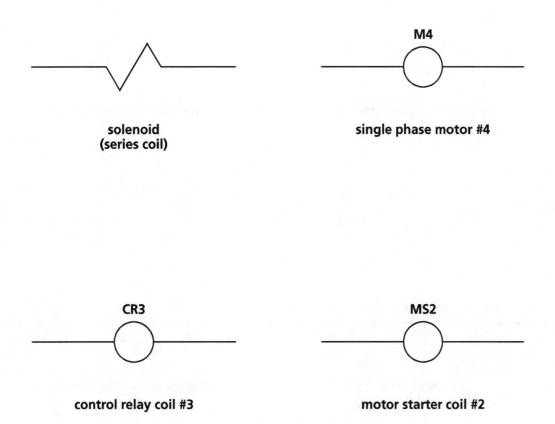

solenoid
(series coil)

single phase motor #4

control relay coil #3

motor starter coil #2

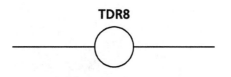

time delay relay #8

Figure 9.

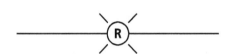

R11

resistor #11 (schematic)

resistor #13 (pictorial)

**red pilot light
(or red indicator light)**

3 phase motor (3ø motor)

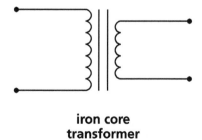

**iron core
transformer**

Overcurrent Protection and Disconnects

Figure 10.

**2 pole, single phase, (1ø)
circuit breaker**

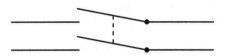

**2 pole, single phase, (1ø)
disconnect**

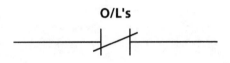

**normally closed
overload relay contacts**

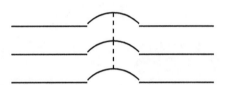

**3 pole, 3 phase, (3ø)
circuit breaker**

**single pole, single phase, (1ø)
circuit breaker**

2 · Safety and ON-OFF Circuits

Figure 11.

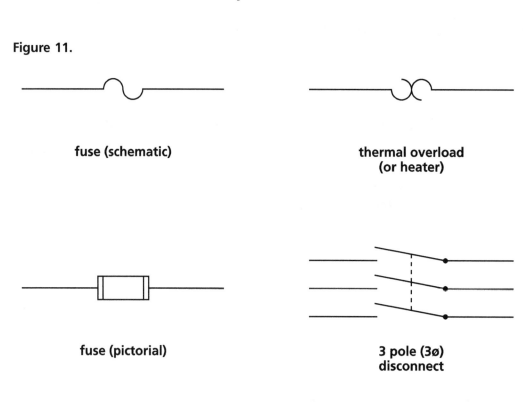

Miscellaneous

Figure 12.

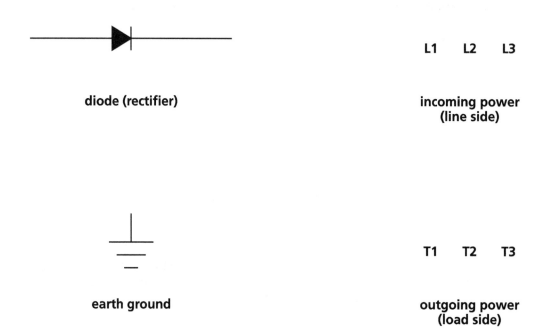

Figure 13.

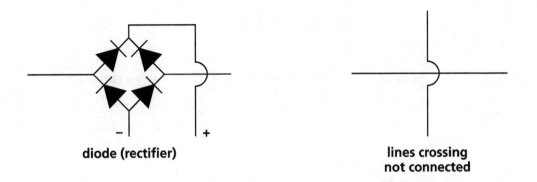

diode (rectifier)   lines crossing
                               not connected

lines connected   battery

mechanically
interconnected

Symbols in the Electronic World

Solid State Diode

Figure 14.

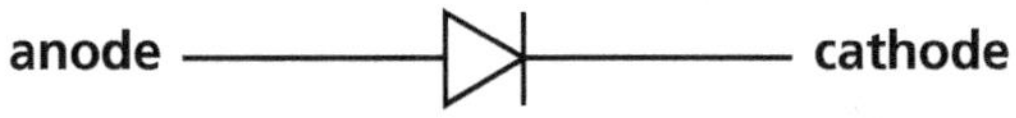

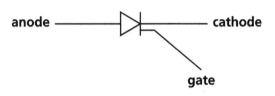

SCR

Figure 15.

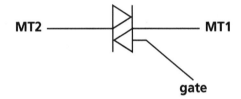

Triac

Figure 16.

LED

Figure 17.

NPN Transistor

Figure 18.

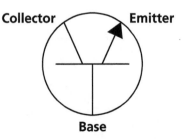

PNP Transistor

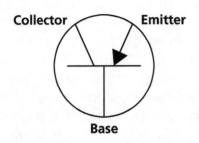

Figure 19.

Regular Resistor

Figure 20.

Pot

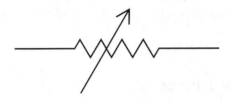

Figure 21.

Photo-Eye

Figure 22.

Chassis Ground

Figure 23.

Earth Ground

Figure 24.

Conductors

There are six general types of named conductors in electricity: hot conductor, ground conductor, neutral conductor, ungrounded conductor, grounded conductor, and grounding conductor. Please note that the 2008 NEC code names are in [square brackets]. The hot wire is called the [ungrounded conductor]. This is a current carrying wire. If you look above your head when at home, this would be either of the two black wires coming into your home. The neutral is called the [neutral conductor] or the [grounded conductor]. This is a current-carrying wire. If you look above your head when at a home, this is the shiny wire holding up the two black wires coming into your home. The ground is called the [grounding conductor] or the [equipment grounding conductor]. This is not a current-carrying conductor. This ground wire exists only to carry fault current and to trip the circuit breaker or fuse on your hot wire circuit when it is drawing too much current.

Double Meanings

Some electrical symbols and terms have double meanings in your work. Some items are called by more than one name, which can be a bit confusing. We are just reporting what you are going to run into. In this book, when we first introduce a symbol or term that has a double meaning or an item with more than one name, we will list the names next to each other.

L1 L2 L3 (power diagrams and schematics)

These are the symbols for the incoming or line side phases of AC electricity in your plant. They are your hot wires on the incoming power side for three phase power. They are also known as phase A, B, and C.

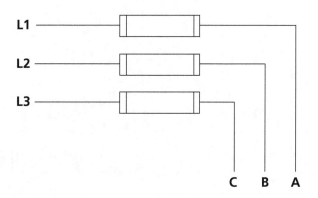

Illustration 2.5
L1 L2 L3 power diagram

L1 L2 (control diagrams and schematics)

L1 is the symbol for the incoming or line side phase of AC electricity in your plant for control wiring. L2 is the symbol for the return line of AC electricity in your plant for control wiring. They are the hot wire and the neutral for single phase power.

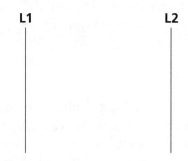

Illustration 2.6
L1 L2 control diagram

T1 T2 T3 (power diagrams and schematics)

These are the symbols for the outgoing or the load side phases of AC electricity in your plant. They are your hot wires on the outgoing power side for three phase power. They are also known as phase A, B, and C.

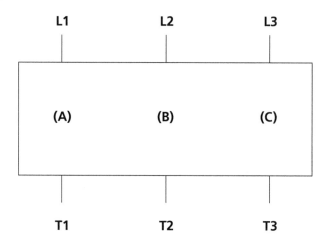

Illustration 2.7
T1 T2 T3 power diagram

Phases

In AC electricity, the term [three phase] means the same as the term three phases and the term [two phase] means the same as the term two phases. In AC electricity, the term [single phase] has two different meanings depending on the context, and does not mean the same thing as the term one phase.

Single Phase

The term [single phase] in AC electricity refers to what the load actually sees (i.e. the actual wave). If you hook a space heater up to 120 VAC, you would have L1 and L2. These are called a hot and a neutral.

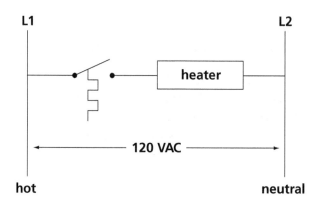

Illustration 2.8
120 VAC load circuit

If you hook a space heater up to 208 VAC, you would have L1 and L2. They are both called a hot.

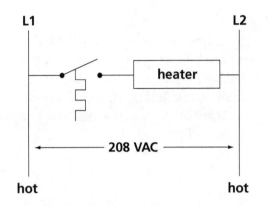

Illustration 2.9
208 VAC load circuit

The 208 VAC circuit has two hot phases, but the load (the heater) sees just one sine wave.

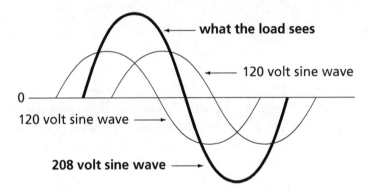

Illustration 2.10
208 VAC sine wave

Exercise 2.1

Build the following circuit using 120 VAC components if you are following along at work or in a school lab setting. Use 24 VAC components if you are building the circuit at home. Use stranded #16 or #14 control wire, a terminal strip, a plug and cord, a pushbutton switch, and a light.

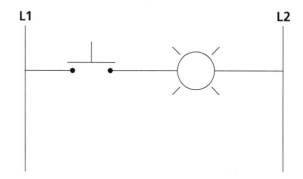

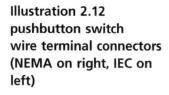

Illustration 2.11
the on off general circuit

Two of the components that you will be using are an industrial type pushbutton switch and an industrial type pilot light. *The wire terminal connectors on the side of electrical devices are generally designed for the insertion of 1 to 2 stripped wires, but no more.*

Illustration 2.12
pushbutton switch
wire terminal connectors
(NEMA on right, IEC on left)

The wire screw connectors on the side of the terminal strip used for the power cord are generally design for the insertion of 1 to 2 *crimped* wires, but no more.

If you put two *crimped* wires under one terminal strip screw, you would place them back to back as shown in Illustration 2.13:

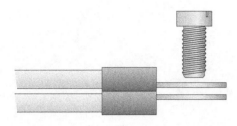

**Illustration 2.13
side view of two
crimped wires under
one terminal strip screw
(yes, OK)**

Not like this:

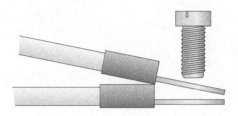

**Illustration 2.14
side view of two
crimped wires under one
terminal strip screw (no,
not OK, bad connection)**

Electrical Troubleshooting and Manual Motor Starter Circuits 3

In most electrical troubleshooting situations, you can do a fairly good job of getting to the bottom of a problem by turning the electricity off and using some sort of a systematic approach and an ohm meter.

Percentages

This system is not the only way to approach troubleshooting, but it works well for us, and we know it can work well for you. A little background on percentages first: if you buy a major electronic device, you will find that about 5 percent of all the new pieces of electronic equipment are dead right out of the box. Even after you have purchased them brand new, about 5 percent are DOA (dead on arrival).

The scenario is similar in the troubleshooting world for major appliances like washers and dryers. After you set up a major appliance in your house, it might not work. So what are you going to do? In most instances, you can call an 800 number and after the initial pleasantries over the telephone line you are going to hear something like this:

> 800# guy: "Mr. or Mrs. Consumer, before we begin in earnest, could you please check for me that the unit is plugged in?"
>
> Consumer: "Mmm, ok. Yep, it is."

800# guy: "OK, let's check that the blue timer is on 35 and the left hand switch is on run."

Consumer: "OK, they are."

800# guy: "Next, let's check the circuit breaker."

Consumer (after a minute or so): "Hey, it was tripped!"

Ta-dahh!

Three things happened here.

A Satisfied Customer

The first success from the company's point of view is that the consumer's product is up and running and the customer is now happy. When we first helped to fix TVs in people's homes, a fair number of the problems were no more than the fact that the set was not plugged in or turned on. You did not just go across the room and plug up the set and relate that fact to the consumer. You had to charge them the company's minimum trip charge, even though you charged for no labor or parts. If you just blurted out the truth then and there, you would quickly get a bruised ego as the consumer would more than likely take out his or her embarrassment on you.

So, what you would do is say something like, "Gee, Mr. or Mrs. Consumer, I could sure use a nice glass of cold water." While they were getting it from the kitchen, you would be bumping around the back of the TV and taking your time to plug up the set. After a few minutes, you would turn the set on and have it playing again. As you wrote out a bill for the trip charge for the visit to their home, you would finish your water. You're happy, the consumer is happy, and the company is happy.

Low Cost

The second success from the company's point of view is that the cost to fix the consumer's product is low. The company knows from experience that the number one thing that is wrong when people call up the 800# line is that the unit is not plugged up. That is why the relatively inexpensive 800# guy reads from a prepared script. The company is playing the percentages to keep from sending out a very expensive repair guy and a service truck.

Quick Turnaround

The third success from the company's point of view is that the consumer's problem was solved quickly.

1-800-Number Troubleshooting

So, how does this relate to a manufacturing plant? In a production situation you want a quick turnaround at a low cost for a satisfied customer. You do not want a 15 minute electrical problem taking a shift and a half to fix. So what we do is to listen to what everyone is saying and see if that is going to solve the problem, but after about 10–15 minutes or so we stop and drop back to some type of system. What follows is what we have observed in the electrical troubleshooting arena.

(1) Is It On or Off?

About 15 percent of the time, the only thing wrong with the machine is that it is turned off, or some part of the machine is turned off. Now of course, if you walk into troubleshoot a machine sight unseen and in 5 minutes or so you flip a switch and the machine starts running, that's good. You are going to have some embarrassed people, however. So, in really volatile situations, we might do something like say, "We think there are coffee and donuts in the break room." At that point in time a lot of the company guys head for the break room as we do the first couple of 1-800# troubleshooting steps.

(2) Fuses and Circuit Breakers

About 15 percent of the time the only thing wrong with the machine is that there is a fuse burned out or a circuit breaker tripped. When you walk over to the machine, the guy that has been working on it might swear to you that the machine has 12 circuit breakers or fuses on it and he has checked every one of them. If you are in system mode however, you ignore what he has told you and check them all yourself. He may have easily overlooked the three fuses in that small fuse box behind the hinge of the door, for example.

(3) Visual Inspection and/or Smell

About 15 percent of the time, the only thing wrong with the machine is that there is a part burnt out (like an overheated solenoid) or a part broken in some fashion (like a shorted terminal). Let us give you an example. We visited a furniture manufacturer that made commercial tops like you might see on cafeteria type tables. A particular machine put a band of material around the edge of the table. It had broken down a few hours before we arrived. After going through steps one and two, we arrived at step three, visual inspection. This particular machine had a cabinet with about twenty or so control relays. When looking at them, we noticed that they all had a light yellow color to them except for two of them, which looked smokey in color.

We asked the one maintenance guy who was still hanging around and not getting some coffee in the break room if the company might have a couple of these control relays in stock. "Sure," he said. He then proceeded to bring two of them back to us. We replaced the two smokey looking control relays with the new ones and then hit the start button on the machine after turning the disconnect back on. Baaaa, ssss, baaaaaaaaaaaa, ssss, baaaa, it went. The machine put a band on the side of the table as it was supposed to do. Elapsed time was about 11½ minutes. Visual inspection was the key.

(4) Bad Terminations

About 15 percent of the time, the only thing wrong with the machine is that there is a loose or broken wire. This happens frequently because it is not unusual for people who learn on the job not to be exposed to certain pieces of key information. Most people in the electrical troubleshooting side of things might know how to use meters, but most of them do not know how to terminate wires properly. In fact, if a company just does a half-day electrical seminar on a Saturday on how to properly terminate wires, within six months or so about 10 percent or more of all their electrical problems will go away. Think of the bang for the buck. It is the most efficient seminar that you can give to your folks. We guarantee it.

(5) Ohm Meter Work

About 5–10 percent of the time, an ohm meter needs to be used to check out something in the circuit or on a component in that circuit. The only thing wrong with the machine is waiting to be discovered with an ohm meter.

The beauty of this particular troubleshooting system is twofold. Firstly, up to this point in time on the 800# list, if you work in an environment where you cannot work around live circuits, that is fine. Up to this point in time you did everything without the power on, using an ohm meter. No volt meter, no PPE, no clamp on ammeter, no live fuses. Secondly, you will still be able to fix about two-thirds of all the electrical circuits that you will ever run into. Think about that for a moment.

(6) Volt Meter and Ammeter Work

About 5–10 percent of the time, the only thing wrong with the machine is that there is something in the circuit or on a component that requires a volt meter or an ammeter. These test instruments need to be used to check out something in the circuit.

(7) The Tough Stuff

The last 15–20 percent of the time it is the tougher stuff, like intermittent problems, VFDs, PLCs, or other things, that need to be fixed. It literally is true that in electrical work, 20 percent of your problems will consume 80 percent of your time and training. You can easily train someone in a motor controls seminar in less than one week's time to fix two-thirds of all the electrical problems that they will ever run into with no power on the machine. A lot of people attempting troubleshooting spin their wheels, but we want you to be successful.

It is sort of like another profession, air conditioning or HVAC. If you pick up any textbook on air conditioning, your impression is that much of what you need to know revolves around gasses, a set of gauges, enthalpy, etc. The truth of the matter in air conditioning work is that if you can do visual inspection, if you can do electrical troubleshooting, and if you can use some common sense, you can fix about 60 percent or more of all air conditioning problems, no gases or gauges needed. Anybody that works in this field can tell you that this is true, but you cannot tell it from looking at an HVAC textbook.

Exercise 3.1

Build the following circuit using 120 VAC components if you are following along at work or in a school lab setting. Use 24 VAC components if you are building the circuit at home. In addition to items from previous circuits, you will need a manual motor starter and a single phase motor.

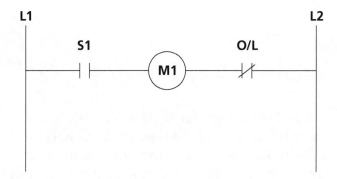

Illustration 3.1
the manual motor starter on off circuit

When attaching wires to the manual motor starter, you will find L1, L2, T1, and T2 at various places on the starter, either in the form of labels or in plastic letters that are actually embossed into the plastic case of the manual motor starter near the terminals. Remember L1 and L2 are the incoming power lines to a device (line side) and T1 and T2 are the outgoing power lines to a device (load side). Your manual motor starter should have them listed on the unit and your motor should have them listed as well.

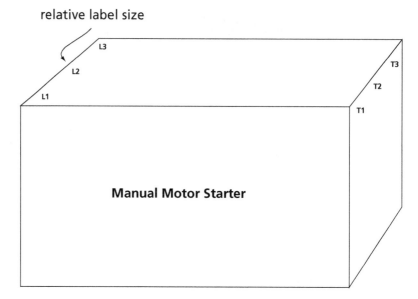

Illustration 3.2
L1, L2, T1, and T2 labels

The main reason for building this circuit is that almost 100 percent of the time in electrical control circuits and in electrical power circuits, the devices on the electrical prints are drawn in the same order as they are hooked up in real life. One of the major exceptions is manual motor contactor control circuits. Don't ask us why. We just report what you are going to run into. In motor circuits and in motor control circuits, what is hooked up out on the machine may not be drawn on the print in the same way or in the same order. You need to keep this in mind when you are troubleshooting these types of circuits.

A good example is the circuit that you built for exercise #3.1. On most machines, the actual hardwired order of the wired up components would be: contact, overload, and motor. As you can see, the print for exercise #3.1 is drawn in this order: contact, motor, and overload. Who knows how this started? We are just the messengers. Please don't shoot us.

Overcurrent Protection and Automatic Control Circuits 4

In the general scheme of things within the world of electricity, you have to protect against contact with the current of a circuit. It is the current and not the voltage that will end up shocking and possibly killing a person, or decimating or destroying a circuit or piece of equipment. Let's look at what happens in two realms, (1) electrical current and people, and (2) electrical current and devices.

Current and People (Touching Hazard)

Almost all of those electrical warning signs that you read are not true. "Danger High Voltage" should really read "Danger High Current." Voltage is a type of electrical pressure similar to the steepness of a stream running by your house. That stream will generally not hurt you because of the low speed and shallow depth of the water that cascades into you as you stand in it.

Current is a type of electrical flow similar to the amount of water that is in the stream running by the front of your house. If that stream floods, the huge amount of water that will spill over you if you stand in the stream bed will probably drown you during a thunderstorm because of the shear volume of water involved. Think garden hose vs. firehouse. Both hoses have the same pressure (about 45 psi), but very different flow rates. Voltage is like the speed and the steepness of a stream, while current is like the flow of a stream.

We can take you to a children's museum. We can hook you up to a super high voltage (200KV) and a very low current (.1 ma) static electricity generator. Your hair will stand on end, your ears may hum, and your lips may crackle if you kiss someone, but your heart will do fine. *The voltage will not get you, it's the current.*

Here is how it works with people on a x10 (times 10) scale:

15 ma

(15 milliamps or .015 amps) is what it takes to light up that red or green LED on your stereo unit. If this amount of current is directly on your heart, then your heart would start to quiver (go into ventricular fibrillation). You would lose consciousness and then die, unless someone was able to get to you, turn the current off, and perform CPR on you in time. Of course, nobody is going to stab you directly in your heart with the two energized leads of an LED, but you can easily get this much current from touching a circuit and getting shocked if you are not careful.

150 ma

(150 milla-amps or .15 amps) is what it takes to pull in the coil of a control relay and/or a contactor. If this much current goes through your muscles, it would cause them to contract and you would not be able to let go of the circuit. You could not let go unless someone turned the circuit off or you were able to drop to your knees and jerk your hand off of the wire. If this amount of current goes directly across your heart, then your heart would stop and you would die.

1.5 amps

(1½ amps or 1500 ma) is what it takes to light up the floodlights in your backyard. If this much current goes through your muscles it will cause them and you

4 · Overcurrent Protection and Automatic Control

to jerk around. It is not a pretty picture. If this amount of current goes directly across your heart, then your heart would stop and you would die.

15 amps

15 amps is what your typical residential circuit is rated at. If this much current goes through your muscles, it will cause them and you to jerk around and burning will start to occur. It is not a pretty picture either. If this amount of current goes directly across your heart, then your heart would stop and cook, and you would die.

The amount of current that your body is exposed to from a "hot" circuit and what effect that it has on you is a function of three things.

1. The *amount and the duration of current* available from the source,
2. the *amount of resistance* of your body (usually how much you are sweating), and
3. *the path* that the current takes through your body.

Current and Devices
(Flash and Explosion Hazard)

If you are wearing your safety glasses that is. If you are not wearing your safety glasses, it will be far worse.

.5 amps

(½ amps or 500 ma) is what it takes to pull in a solenoid in your typical commercial or industrial circuit. If this much current shorts out it, could cause you to jerk back and fall off a ladder or hit something the wrong way and cut yourself.

5 amps

(5 amperes) is what it takes to run a ¾ horsepower motor on your typical commercial or residential 120 VAC circuit, such as a big whole house fan or a pump.

If this much current shorts out, it could cause you to jerk back and cut yourself or fall off a ladder. You may hit something very hard the wrong way and injure yourself, but the current flash itself will not harm you.

50 amps

(50 amperes) is what it takes to run an electric stove on your typical commercial or residential 240 VAC circuit. If this much current shorts out, you might think that you are in the middle of a welding situation. You could be temporarily blinded or could hurt yourself further in some sort of industrial accident.

500 amps

(500 amperes) is what it takes to crank a motor over on your typical industrial, commercial, or residential vehicle, such as a car or truck. If this much current shorts out, it could cause you to be blinded, jerk back and fall, hit something the wrong way and stab yourself by accident, or be burned by large sparks.

5,000 amps

(5,000 amperes) is what it takes to run a typical industrial production line, or a commercial building. If this much current shorts out it could cause you to suffer an explosion and blast. It may cause you to jerk back and fall off a ladder and forcible hit your head. You might also run into something the wrong way like a forklift and injure yourself very badly. You would also be blinded or burned by flying molten metal.

50,000 amps

(50,000 amperes) is a major explosion. If this much current shorts out it, could easily cause you to die.

How do you protect yourself from these hazards? There often is a debate in the electrical trade as to the

4 · Overcurrent Protection and Automatic Control

order of the following items, but not on the actual list of items itself.

Residential and Office Space Current Safety Considerations

Leave all ground wires and ground plugs intact. This is to insure that the circuit breaker will trip if you have a short. This is used primarily to protect people and equipment.

Use heavy duty extension cords. This is to make sure that the extension cord will not overheat and cause a fire to break out. This is used primarily to protect people and equipment.

Make sure all equipment is working properly (switches not broken, cover plates on, etc.). This is used primarily to protect people.

Make sure all AFCI [arc fault circuit interrupter] circuit breakers are working. This is to make sure that the breaker will trip if you have electrical arcing in your circuit. This is used primarily to protect the building and equipment.

Make sure all GFCI [ground fault circuit interrupter] outlets or circuit breakers are working. This is to make sure that the breaker will trip if you have a leakage current in your circuit (usually around liquids or moisture). This is used primarily to protect people.

Industrial Current Safety Considerations

Make sure that all grounds, wires, conduits, bushings, etc., remain intact. This is to make sure that the circuit breaker will have a good ground path and will trip if you have a short. This is used primarily to protect people and equipment.

Throw out any cut or damaged extension cords. This is to make sure that the extension cord will not shock someone. This is used primarily to protect people.

Make sure all circuit breakers and fuses are in working condition (i.e. the fuses are the correct size, both in their electrical current value and their clearing time value and the circuit breakers are mounted or bolted on tightly and have the right tie handles in place). This is to make sure that the breaker will trip or that the fuse will blow if you have a short in your circuit. This is used primarily to protect people (with fuses) and the building and equipment (with circuit breakers).

Turn off the power if you are going to troubleshoot electrical circuits. You can still fix two-thirds of the stuff anyway with no power on. That's still going to be a lot better than a majority of the folks working on electrical circuits out there are doing. This is used primarily to protect people.

Use LOTO (lock out tag out). This is used primarily to protect people.

If you need to work on or around live electrical circuits use additional PPE (personal protective equipment) as referenced in NFPA 70E.

This is used primarily to protect people.

Industrial Overcurrent Protection and Time

To protect people and equipment from electrical hazards one must be aware of time, usually in the form of milliseconds (1/1000 of a second), cycles (1/60 of a second), seconds, and minutes. As in real life, time can seem compressed (for example, shooting at targets for 7 minutes) or it can seem very long (serving time in prison for 3 years). In order to protect people and equipment within different time realms, different overcurrent protection is generally used.

For very, very short time frames on the order of milliseconds, solid state circuits like crowbar circuits are used. For very short time frames on the order of a ¼ cycle or several cycles, fused circuits are used. For short time frames on the order of fractional cycles or fractional seconds, circuit breaker circuits are used. For short time frames on the order of seconds or minutes, motor overload protection circuits are used.

In order to prevent production line shutdowns due to false triggering in an industrial situation, the following overcurrent protections are the most popular (generally in this order): circuit breakers, fuses, motor overload protection, and solid state devices.

Automatic Control Circuits

Automatic control circuits are also known as two wire control circuits, because if you would cut the cable or the conduit between the sensor and the load, it would have two conducting wires in it.

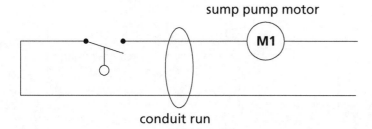

Illustration 4.1
two wire control circuit

As you can see in this circuit, the sump pump is hooked to a float switch. When the water inside the home's basement rises too high, the float switch closes and the pump turns on. It runs until the pump lowers the water level in the basement of the house and the float switch turns the pump off. The major downside with this type of circuit is that if the power goes off for some reason, you would be in a bad position. You would not know for sure whether or not the pump would turn on immediately when the power was restored.

For example, someone in a plant may be working on a ventilator fan at the top of a 12-foot extension ladder, while the power is off. He is wearing a harness, but because there are no hook-up points near him, he is unattached. During the time that he is working on the fan, the temperature rises and the thermostat makes (turns on), but since the power is off the fan does not start. It is usually too noisy to hear the thermostat click on in a plant. At some point in time, the power is switched back on and the fan starts up and knocks him off the ladder. Statistics show that in the U.S., half the people that fall from a ladder of 11 ft. or more die.

Exercise 4.1

Build the following circuit using 120 VAC components if you are following along at work or in a school lab setting. Use 24 VAC components if you are building the circuit at home. In addition to items from previous circuits, you will need a pushbutton switch or a limit switch to mimic the action of a thermostat.

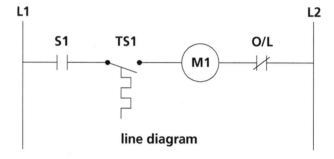

line diagram

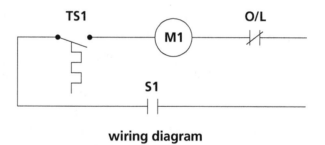

wiring diagram

**Illustration 4.2
the automatic motor starter two wire control circuit**

Ohm Meter Testing 5

As we have stated before, about two-thirds of all electrical problems can be fixed with the power off. One of the instruments that you will use is the ohm meter. The ohm meter is used on a circuit to test for one of five things: continuity, open, short, leakage, or value. In this chapter, we will be using a digital VOM as an ohm meter, since that is the most common type of meter used on the job.

Continuity

To check for continuity in most instances, simply unplug or disconnect the device from the circuit. Turn the instrument on and put it on the sound or the continuity setting of the digital meter. Touch the two meter probes together to make sure that the meter works, then touch the two probes (usually black and red) to the two ends of the device being tested. If the device is not burnt out, the meter beeps and you pass (PASS), which means that the device is OK. If the device is burnt out, the meter doesn't beep and you do not pass (FAIL), which means that the device is BAD.

The only problem with continuity testing is that on average you will probably miss about 15 percent of your electrical problems. In order to be more precise, you would have to do what we call the solenoid or coil test (open, short, leakage, and value) with your ohm meter. This would be done in order to catch intermittent problems and more advanced problems, like a three phase motor that is about to fail.

Solenoid Test

Open

An open occurs in a solenoid or a coil circuit when it burns in two. To check for an open in most instances, unplug or disconnect the device from the electrical circuit. Then turn the meter on, click over to ohms, and put it on the R×10,000 scale or on the automatic scale setting of the digital meter. Touch the two meter probes together to make sure that the meter works, then touch the two probes (usually black and red) to the two ends of the device being tested.

If the device is good, the meter reads some value and you pass (PASS). If the device is burnt out, the meter reads infinity (sometimes OL or sideways eight—∞) and you do not pass (FAIL), which means that the device or the wiring is BAD. *Make sure that your fingers do not touch the metal probe tips of the ohm meter, as this will give you a false reading.*

Short

A short occurs in a solenoid circuit when it basically shorts out in the two wires going to the coil or in the two wires just inside the coil. To check for a short in most instances, unplug or disconnect the device from the circuit. You turn the meter on, click over to ohms, and put it on the R×1 scale of the digital meter or put it on the meter's automatic scale setting (the same one used in the open test). Make sure that the meter works by touching the probes together and then touch the two probes (usually black and red) to the two ends of the device being tested. The device is good if the meter reads some value (but not zero) and you pass (PASS), which means that the device is OK. If the device is shorted out the meter reads zero (sometimes exactly 0 or sometimes with digital meters it reads really close to zero, like .001 ohms) and you do not pass (FAIL), which means that the device or the wiring is shorted and the device is BAD.

Value

While you are doing the short test, you can also write down the actual value of the coil in ohms. In order for you to check for a value in most instances, unplug or disconnect the device from the circuit, turn the meter on, click over to ohms, and put it on the R×10 or the R×100 scale or the automatic scale setting of the digital meter. Make sure that the meter works by touching the two probes and then touch the two probes (usually black and red) to the two ends of the device being tested. If the meter reads some value and a replacement part that you have in stock has the same value or a very similar value then you pass (PASS). This means that the device is OK. If the device is not reading a value close to the replacement part's value, then you do not pass (FAIL), which means that the device or the wiring is BAD.

Leakage

Leakage occurs in a solenoid circuit when it starts shorting out to ground. To check for leakage in most instances, unplug or disconnect the device from the circuit and turn the meter on, click over to ohms, and put it on the R×10,000 scale or the automatic scale setting of the digital meter. Make sure that the meter works and then touch one of the two probes (either black or red) to the one of the ends of the device being tested and the other end to the case or the ground of the machine being tested. If the device is not leaking electrons, then there is no path to ground through the coil. The meter reads infinity (sideways eight or sometimes OL) and you pass (PASS). This means that the device or the wiring is not leaking electrons and the device is GOOD. If the meter reads some value, you do not pass (FAIL), which means that the device is BAD.

Test for Switches and Relay Contacts

Use the continuity test for a quick check. Flip the switch on and off several times to see which of three states show up in the switch contact being tested. The digital meter should beep and then not beep, and then beep again if you are GOOD, when flipping the switch on and off. The digital meter should beep and then keep beeping if you are BAD, when flipping the switch on and off if the switch contact is always closed. The digital meter should not beep and then never beep if you are BAD, when flipping the switch on and off if the switch contact is always open.

Use the value test for a more advanced test. The switch is GOOD if you read infinity in the first switch position and read less then five ohms in the other switch position, otherwise the switch or contact is BAD or will fail in a very short order. In summary:

First position reading	Second position reading	Test result
infinity	<5 ohms	good
infinity	>5 ohms	bad
infinity	infinity	bad
0 ohms	0 ohms	bad
<5 ohms	infinity	good
>5 ohms	infinity	bad
>5 ohms	>5 ohms	bad

Practice Makes Perfect

In order to get a get a sense of how these tests work, we will do a little practice testing. If you already know how to test for bad fuses then skip this practice testing section, but make sure that you do exercise #5.1.

At your workplace or at your desk, gather as many orphaned and discarded fuses as you can possibly find and throw them all into a pile or into a box. Get your digital meter and set it up for continuity testing. Turn the meter on and put it on the sound or the continuity setting mode. Touch the two meter probes together to make sure that the meter works. Then, touch the two probes (usually black and red) to the two ends of the fuse being tested. If the fuse is good the meter beeps and you pass (PASS), which means that the fuse is OK. If the fuse is burnt out the meter will not beep and you do not pass (FAIL), which means that the fuse is BAD.

Go through the whole pile of fuses and separate the bad ones from the good ones. In most situations, you should have more bad ones than good ones. Beep, beep.

Now, gather as many orphaned and discarded solenoids, heaters, motors, and muffin fans as you can possible find (anything that has a coil of wire in it) and throw them all into a pile or into a box. Get your digital meter and set it up for solenoid testing. Turn the meter on, set it on ohms, and put it on the automatic scale setting. Touch the two meter probes together to make sure that the meter works. Then, touch the two probes (usually black and red) to the two ends of the device being tested. Test for open, short, and leakage. Please note that in this practice drill we are not testing for value because you will not have a replacement coil available for most of the items that were pulled out of the trash.

If the coil is good the meter reads a value and you pass (PASS), which means that the coil is OK. If the coil is burnt out, the meter doesn't read, reads zero, or reads very close to zero, then stop. You do not pass (FAIL), which means that the coil is BAD.

Turn the meter on and put it on the automatic scale setting. Touch the two meter probes together to make sure that the meter works. Then, touch the two probes (usually black and red) to the two ends of the device being tested. Test for leakage. Touch one of the two probes (either black or red) to one of the ends of the coil being tested and the other end to the case or the ground of the coil being tested. If the coil is not leaking electrons, then there is no path to ground through the coil. The meter reads infinity (sometimes OL or sideways eight) you pass (PASS). This means that the coil or the wiring of the coil is not leaking electrons and the device is GOOD. If the meter reads some value then you do not pass (FAIL), which means that the device is BAD.

Exercise 5.1

Do the following exercise using a 230 VAC/460 VAC dual voltage nine lead, three phase motor, whether you are doing it at work, in a school lab setting, or at home. It doesn't matter if the motor is good or not. Use a three phase motor obtained from a junkyard, if you are doing this exercise at home. You can generally buy one inexpensively for the scrap value or, with permission, you can dumpster dive at a plant for one from the metal recycling bin or the maintenance shop dumpster. In addition to items from previous circuits, you will need a digital VOM.

You will be testing for five things: mechanical operation, open, short, leakage, and value. You will be testing one thing with your hand and four things with your meter.

Mechanical Operation

Grab the shaft of the motor and rotate. If the bearings seem smooth, you PASS. Go on to the next test. If not, then you FAIL.

Grab the shaft of the motor and try to move the shaft at right angles to the rotation in order to see if the bearings have any sideway motion. If you have no motion at right angles, you PASS. Go on to the next test.

If not, then you FAIL. No need to do any more testing. Stop here.

Modified Solenoid Test

Hook the motor up with wire nuts for the high voltage Wye or the high voltage delta connection T7 to T4, T8 to T5, T9 to T6 (that way you do not need to know if the motor is a Wye or a delta). The testing will then be done on T1, T2, and T3.

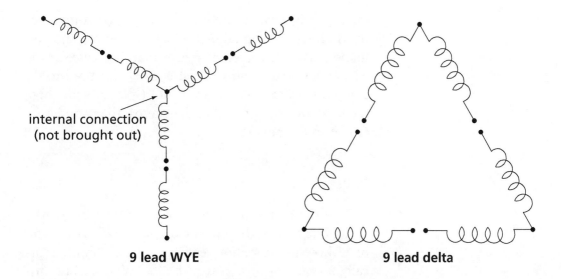

9 lead WYE   **9 lead delta**

Illustration 5.1
wye and delta
configuration

Open

Get your digital VOM meter and set it up for ohm testing. Turn the meter on and put it on the automatic scale setting or the R×10000 scale. Touch the two meter probes together to make sure that the meter works. Then, touch the two probes (usually black and red) to the two ends of the coil being tested. In this case you have three coils and hence three readings to take, T1 and T2, T1 and T3, and T2 and T3. If any of the three coils of the three phase motor are burnt in two (burnt out) the meter will read infinity (sometimes OL or sideways eight) and you do not pass (FAIL). This means that the coil wiring is open and the device is BAD. No need to do any more testing. Stop here. If not (if your three readings are anything other then infinity) then PASS. Go on to the next test.

Short

Put the meter on the R×1 scale or on the automatic scale setting of the digital VOM meter on the ohms setting. Touch the two meter probes together to make sure that the meter still works. Then, touch the two probes (usually black and red) to the two ends of the coils being tested. If the device is shorted out, the meter reads zero (sometimes exactly 0 or sometimes really close on a digital meter like .001 ohms) and you do not pass (FAIL), which means that the device or the wiring is shorted and the coil is BAD. No need to do any more testing. Stop here. If not (if all of your three readings are anything other than zero), then you PASS. Go on to the next test.

Value

Take the meter and put it on the R×1 scale or the automatic scale setting of the digital meter, as the coil of a three-phase motor reads very small ohms, typically 5 ohms or less. Touch the two meter probes together to make sure that the meter still works. Then, touch the two probes (usually black and red) to the two ends of the coil being tested. The meter will read some value. Write down the three values that you obtain. The values are not important—however, all three must be within 5 percent of each other. A three-phase motor is built symmetrically. If the coils are not reading a value close to each other then you do not pass (FAIL), which means that the device or the wiring is BAD. No need to do any more testing. Stop here. If the three readings are the same value or a very close value (within 5 percent or less of each other), then you pass (PASS), which means that the coil values are OK. Go on to the next test.

Leakage

Leakage occurs in a coil circuit when it starts shorting out to ground (to the metal case of the motor, in this case). Put the meter it on the R×10000 scale or the automatic scale setting. Touch the two meter probes together to make sure that the ohm meter still works. Then, touch one of the two probes to one of the ends of the coil being tested and the other end to the case or the ground of the motor being tested. Be sure and scratch through the paint and get down to the metal on the side or the foot of the motor in order to get a good reading. Do this for all three coils. If the meter reads some value, you do not pass (FAIL), which means that the device is BAD. No need to do any more testing. Stop here. The motor is bad. If the device is not leaking electrons, then there is no path to ground through the coil. If the meter reads infinity three times then you pass (PASS). This means that the coils or the wiring are not leaking electrons and the coils are GOOD. No need to go on to the next test, this is it. The three phase motor is good.

You can use this test to check a new motor, an old dusty motor in stock, or one out on a machine still hooked up. If you need to check out a three phase motor on a machine, just turn off the power and check it out with the meter at the bottom of the motor's motor starter at T1, T2, and T3.

Start-Stop Pushbutton Circuits (Latch-Unlatch) 6

The most visible and one of the most common types of control circuit used in industry is the start-stop pushbutton circuit.

Illustration 6.1
start stop motor control circuit

Any machine that has a motor in it with pushbutton switches has this circuit. This circuit is known by several other names.

Three Wire Control Circuit

This name came about for the start-stop pushbutton circuit because if you took an axe and chopped the circuit in half at the conduit run, you would have three wires inside.

Illustration 6.2
start stop motor control circuit wiring diagram (two stations shown)

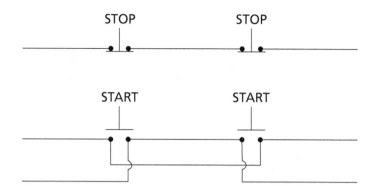

Each conduit that runs between each pushbutton station would have three wires in it.

No and Low Voltage Protection Circuit

This name came about for the start-stop pushbutton circuit because if you lose power during a machine's operation, the circuit will drop out the motor starter's coil and it will not re-engage (turn on) the motor when the power comes back on. This makes the start-stop pushbutton circuit a very safe one, compared to a two-wire or an automatic control circuit.

Latch-Unlatch (Latching)

This name came about for the start-stop pushbutton circuit because in other industries (electronics, telephones, computers, and PLCs), this circuit has always been known as a latch-unlatch circuit or as a latching circuit.

Illustration 6.3
latch unlatch, start stop motor control circuit PLC diagram (traditional, as in maintenance mechanics and electrical control techs)

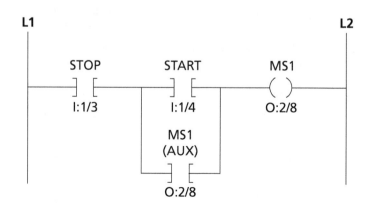

6 · Start-Stop Pushbutton Circuits (Latch-Unlatch)

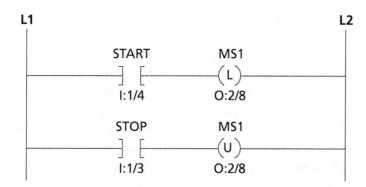

Illustration 6.4
latch unlatch, start stop motor control circuit PLC diagram (modern, as in electronic and PLC techs)

Either approach works. The only difference between the approaches is the thought process used in the electronic world vs. the thought process used in the electrical world. The bottom line is that (1) you want the circuit to work, (2) you want the circuit to be safe, and (3) you want to document any changes that you make to the circuit.

Let's examine the basic way that this circuit works and how it protects a machine operator. If you look at the flow of the circuit (illustration 6.5 from left to right), you will see that the circuit shows the rest position of the machine. The power is at L1 on the incoming rail and goes through the normally closed contact of the stop pushbutton PB1. The power can not go any further than the left hand side of the pushbutton contact PB2 or the left hand side of the auxiliary contact C1 because they are both normally open (N.O.) in the rest position.

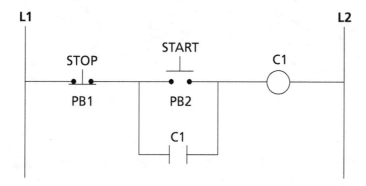

Illustration 6.5
a latching contactor circuit

The N.O. auxiliary contact C1 and the contactor coil C1 both have the same label because in the electrical world both items are on the same device (contactor C1). When the coil changes state (from energized to non-energized or vice versa) then the contact will change state (from N.C. to N.O. or vice versa).

The next step in the sequence of operation is that someone pushes the start pushbutton PB2 and the current flows through to coil C1, and then through that coil to the power rail L2.

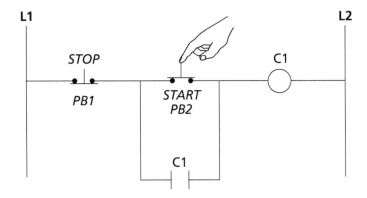

Illustration 6.6
the finger of fate pushes START

In the next fraction of a second, coil C1 is energized and the N.O. contact C1 closes.

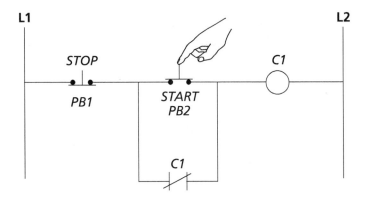

Illustration 6.7
the coil energizes

Now, there are two paths for the current to get to the coil C1. One is through the pushbutton PB2, which still has the finger on it, and the other path is through the contact C1. The circuit is said to be latched.

In the next instance, when the finger is being withdrawn off of pushbutton PB2, the contact C1 remains closed and the coil C1 remains energized.

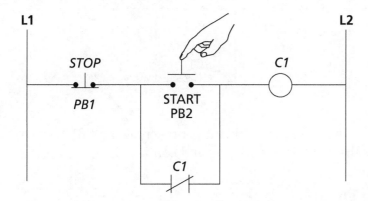

**Illustration 6.8
the pushbutton bounces back (it is spring loaded)**

Power continues to flow from L1 through the N.C. PB1 contact, through closed contact C1, through coil C1, and finally to L2.

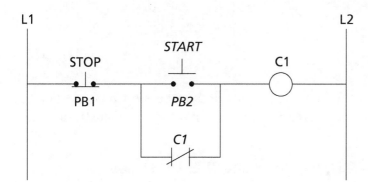

**Illustration 6.9
the energized circuit**

Please note that *electrical circuits are generally drawn in the de-energized state.* This means that on a print, you will see the circuit drawn as it is laid out in illustration 6.5. You will not see the circuit drawn in its energized state (illustration 6.9) on a print. That is why some of the circuits that we drew (illustrations 6.6 to 6.9 and others) are labeled in a different fashion in this book. The labels between incoming and outgoing rails L1 and L2 are drawn with italic print if the circuit is illustrated in its energized state.

The way to stop this circuit (i.e., to de-energize or unlatch it) is to push the stop button (PB1).

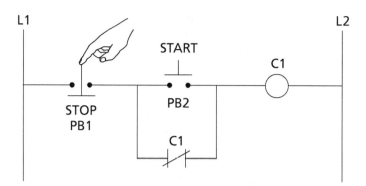

Illustration 6.10 stop in the name of ...

As soon as the circuit is broken at the PB1 contact, the coil C1 de-energizes and contact C1 opens up.

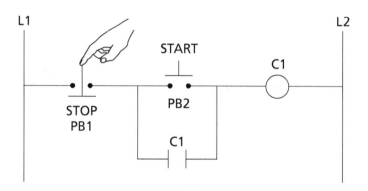

Illustration 6.11 still being pushed off, a fraction of a second later

When the finger releases the stop pushbutton, the circuit returns to its original state (see illustration 6.5).

N.O. vs. Open and N.C. vs. Closed

In these switching circuits, *the words normally open (N.O.) and normally closed (N.C.) refer to the state of a device in its rest position or its de-energized state. The words open and closed refer to the state of a device in its activated position or its energized state.*

Open and closed in electrical work mean the complete opposite of the words open and closed in just about any other profession. In just about any other trade in the world, if you close a valve, close a gate, or close a door, you shut off a flow. Not in electricity. In just about any other trade, if you open a valve, open a gate, or open a door you start a flow. Again, not in electricity.

6 · Start-Stop Pushbutton Circuits (Latch-Unlatch)

This is a normally open pushbutton switch that is open.

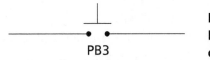

Illustration 6.12
N.O. pushbutton that is open

This is the same pushbutton switch being pressed. It is still hooked up N.O. and it is still called N.O., but now it is in the closed position.

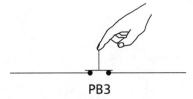

Illustration 6.13
N.O. pushbutton that is closed

Another way to think of this is a light switch, like one in your home. It is a single pole switch. It is a N.O. switch. In the rest (off) position, the switch is open and the lights in the room are off. In the activated (on) position the switch is closed and the lights in the room are on.

Exercise 6.1

Build the following circuit using 120 VAC components if you are following along at work or in a school lab setting. Use 24 VAC components if you are building the circuit at home. In addition to items from previous circuits, you will need a contactor.

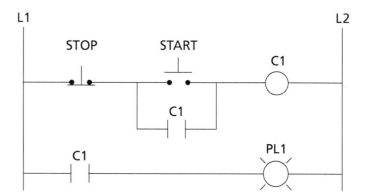

Illustration 6.14 latching contactor circuit using a pilot light as the load

The name of the relay used in a motor control circuit is contingent on how many amps of current it switches. The naming convention looks something like this:

Reed Relays

Reed relays are used in low voltage and very low current switching applications (typically 5–24 VDC and up to 24 VAC). They are used on doorways in burglar alarm systems, and on hatches and doors of production equipment. They are usually installed across from a magnet. The magnet causes the contact of the Reed relay to open and close as the door opens and closes.

Control Relays

Control relays are used in low voltage, low current (non-motor or non-light bank) switching applications (typically 6–120 VDC and 12–240 VAC, and up to 10 amps). They are used for control flexibility and to have one voltage level (like 24 VAC from a sensor) to be able to switch a different voltage level (like a contactor coil rated at 120 VAC) in your control circuit.

Power Relays

Power relays are used in low and regular voltage, higher current (non-motor or non-light bank) switching applications (typically 6–120 VDC and 12–240 VAC, and up to 25 amps). They are used for control flexibility and to have one voltage level (i.e. 24 VAC from a HVAC sensor) switch a different voltage level (i.e. 120 VAC) in a contactor coil.

Contactors

Contactors are relays that are used in regular voltage, higher current (motor and/or light bank) switching applications (typically 120–240 VAC and up to 50 amps or so). They are used for control flexibility and to have one voltage level (i.e. 24 VAC from a HVAC sensor) to be able to switch a heavy current load on and off (such as a 238 VAC compressor).

Motor Starters

Motor starters are contactors with overload protection that are used in regular voltage, higher current (motor and/or light bank) switching applications (typically 120–240 VAC and up to 500 amps or so). They are used for control flexibility and to have one voltage level (such as 120 VAC from a pushbutton switch) able to switch a heavy current load on and off (such as a 600 VAC three phase motor).

Solid State Switches 7

In addition to standard options like pushbutton switches and limit switches, there are quite a few solid-state (non-mechanical) switching options available in industry. The most common types are proximity detectors and photoeyes.

Proximity Detectors

Proximity detectors are designed to pick up a target at close range and give a signal to your control circuit as to whether or not something moves in front of the sensor or passes through a gate near the sensor, etc. The proximity detectors basically induce a small magnetic field out in the environment close to the prox (proximity detector). The proximity detectors are then set up to detect the imbalance of this field caused by the closeness of a target. The inductive type of proximity detector is set up to use a target of metal, such as the side of a metal production gate etc. This feeds movement information back into a control circuit.

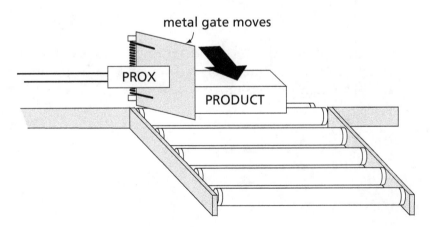

**Illustration 7.1
inductive prox. targeting**

The capacitive type of proximity detector is set up to use the product itself as the target. That feeds movement information through capacitive interaction with the electric field back into a control circuit.

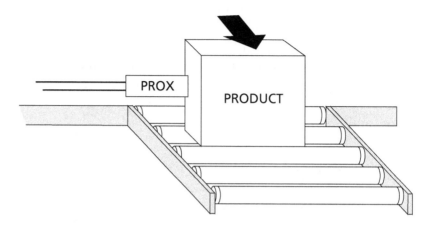

Illustration 7.2 capacitive prox. targeting

Proximity detectors look almost exactly like a switch in a control circuit and they come in both N.O. and N.C. versions.

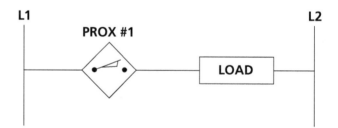

Illustration 7.3 AC prox. switching circuit

The two wire version tends to be the most popular AC one. The three wire version tends to be the most common DC version.

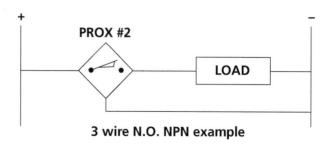

3 wire N.O. NPN example

Illustration 7.4 three wire DC prox. circuit

The great thing about using a prox is that the two wire version can be dropped right into a circuit to replace a mechanical switch. The proximity detector switches are already sealed with no moving parts. This means that they can be used in a somewhat unfriendly environment. They will not short out if moisture gets on the outside of the switch. The cost per unit in an industrial environment is about equal to the cost of an industrial switch. There is also a four wire version that lets you choose either N.O. or N.C. on the same prox. sensor.

The sensors also come in shielded and un-shielded versions. For example, if you want to have a N.O. prox. sensor looking through a hole in the side of a conveyor at your product going by and want the sensor face to be flush with the side of the conveyor, then use the shielded version. If you use the un-shielded version of the proximity sensor in this same situation, it would always pick up the side of the conveyor and always be ON or closed.

Photoeyes

The other common way to do non-contact switching in an industrial setting is to use photo electric devices (photoeyes). The cost in an industrial environment for photoeyes is about two or three times the cost of an installed electric switch, but they are a lot more flexible. The most common types are retro reflective, diffuse, thru beam, and convergent.

Most photoeyes are made of pulsing LED lights on the emitter side of the circuit and a tuned receiver on the detector side of the circuit. This prevents false triggering. In the old days of photoeye usage, the emitter was just a regular light source with either a mirror or a receiver on the other side of the conveyor. In either case, if you used a flashlight for working on something on the machine, you could trigger the photoeye by mistake. Today, however, the emitter may pulse at 40 kHz or so and the detector is tuned to detect at that frequency. If you have a 60 Hz overhead light or a 0 Hz flashlight near the detector, it will not pick it up, and give you false triggering.

Retro Reflective

If the emitter and the receiver are in one unit that looks out over a conveyor or some other space to a mirrored tape or some other type of reflector, the photo eye is called retro reflective. The mirrored tape or the state-of-the-art reflector is designed to directly bounce back to the source most of the incoming signal using microscopic versions of the corner cube reflectors or spherical cats eyes (in the form of tiny glass beads for heaven's sake).

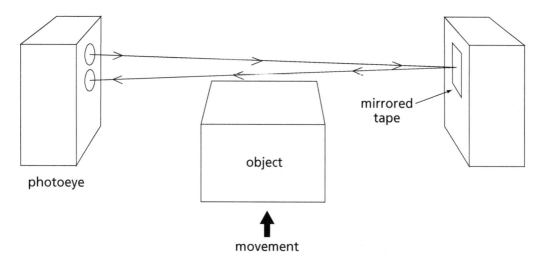

**Illustration 7.5
retro reflective photoeye**

In this setup, you do not have to constantly adjust the angle of a mirror to exactly bounce back most of the signal. A real pain in the ... in the old days of photoeyes. If you do use a regular mirror for this type of setup, it would be called a reflective photoeye instead of a retro reflective one.

Diffuse

If the emitter and the receiver are in one unit that looks out over a conveyor or some other space, and it looks for the object itself to reflect back the light signal, then the photoeye is called diffuse (or a proximity photoeye). The photoeye unit is designed to directly bounce back to the receiver part of the incoming signal from the emitter. Needless to say, the photoeye has to be very close to the product that it is designed to pick up when you are in this mode because the product or the package itself absorbs much of the incoming light beam.

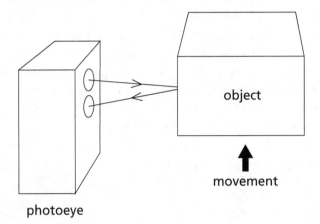

**Illustration 7.6
diffuse photoeye**

Thru Beam

If the emitter and the receiver are in two separate units and the emitter looks out over a conveyor or some other space to a receiver, the photo eye is called thru beam. The two units are designed to detect light directly from the source (the emitter) shining in a straight line directly into the receiver. This method is used to enhance the distance covered by the incoming signal.

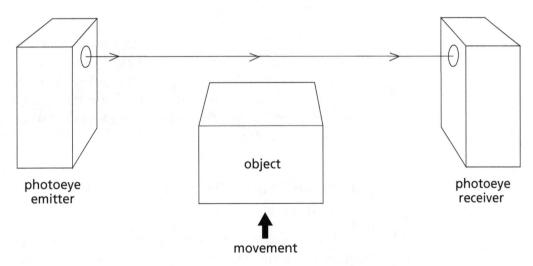

**Illustration 7.7
thru beam photoeye**

Convergent

If the emitter and the receiver are in one unit that looks out over some small space to a small object or some small appendage of that object, then the photo eye is called convergent. The signal is designed to focus (through the use of a lens) directly onto a very small circle. This is done in order to pick up a very small part (like the legs of an IC chip or a thread in a weaving machine). The appendage of the object then bounces some of the incoming signal back to the receiver. This is similar in operation to the diffuse photoeye, but this beam is focused to a very small point.

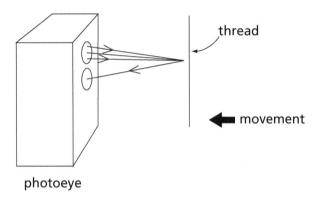

Illustration 7.8 convergent photoeye

Light Dark Operate

This is like having a reverse switch on your photoeye circuit. If the photoeye circuit turns ON when it detects an object and you want it to turn OFF when you detect an object, then toggle the light-dark operate switch. Then, bingo! The circuit now turns OFF when you detect an object.

Multiple Hookup Details of Solid State Switching

Two wire devices tend to have much higher leakage currents when they are in the OFF mode than three wire devices. If you need to hook up two or more proximity detectors or photoeyes in series (AND) or parallel (OR), generally you would use three wire devices. When you need to hook up proximity detectors or photoeyes to a PLC, use three wire devices or two wire devices with some type of buffer, like a pull-up or pull-down resistor.

Exercise 7.1

Build the following circuit using 120 VAC components if you are following along at work or in a school lab setting. Use 24 VAC components if you are building the circuit at home. In addition to items from previous circuits, you will need a motor starter.

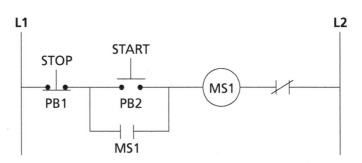

Illustration 7.9
start stop motor starter circuit using a contactor and an overload relay as a motor starter

Exercise 7.2

Build the following circuit using 120 VAC components if you are following along at work or in a school lab setting. Use 24 VAC components if you are building the circuit at home. In addition to items from previous circuits, you will need an AC prox. detector.

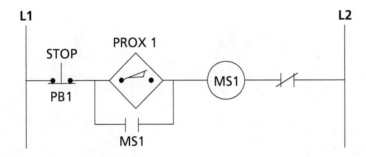

Illustration 7.10 start stop motor starter circuit using a motor starter and an AC prox. switch

Exercise 7.3

Build the following circuit using 120 VAC components if you are following along at work or in a school lab setting. Use 24 VAC components if you are building the circuit at home. In addition to items from previous circuits, you will need a motor.

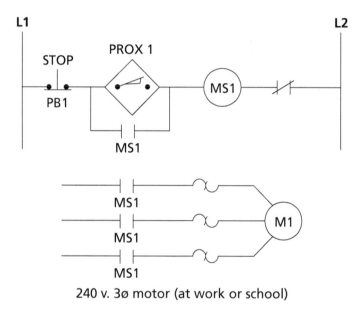

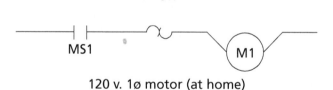

Illustration 7.11 start stop motor starter circuit using a motor starter and a three phase or a single phase motor as the load

Wye and Delta 8

In addition to ohms, current, and voltages, in electrical work there are standard relationships and hookups. There are math relationships (the square root of 2 and the square root of 3) and standard industrial hookups (Wye and delta) for three phase transformers and three phase motors.

Three Phase Transformers

When dealing with industrial plants, the most common types of transformer hookup by far are the three phase Wye and the three phase delta hookup.

Wye

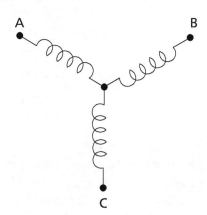

Illustration 8.1
three phase Wye

Voltages with Wye transformer hookups are related to 1.732 (the square root of 3). For example, if you have a three phase Wye transformer secondary hookup that is 480 VAC and pull a neutral [neutral conductor] off the center point [neutral point] you will get 277 VAC phase to neutral.

99

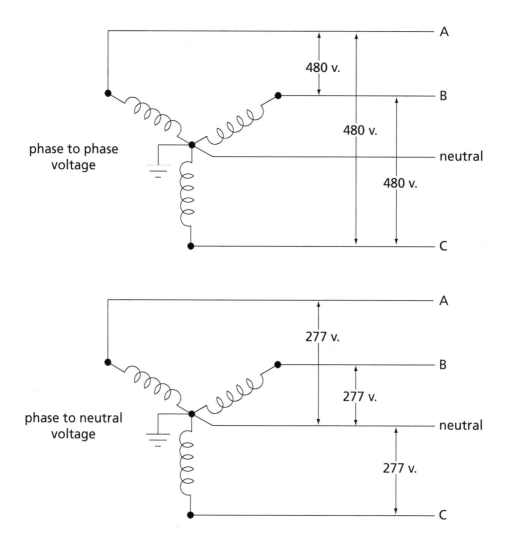

**Illustration 8.2
480 VAC three phase Wye, 277 VAC single phase**

This is a commonly used transformer hookup if you need single phase voltage for the overhead lights in an industrial plant warehouse. The single pole breakers that protect most of the overhead lights in that building are 277 VAC single phase. Again, with Wye transformer hookups, voltages are related to 1.732 (the square root of 3).

If you have a three phase Wye transformer secondary hookup that is 208 VAC phase to phase and pull a neutral off the center point, you will get 120 VAC phase to neutral.

8 · Wye and Delta

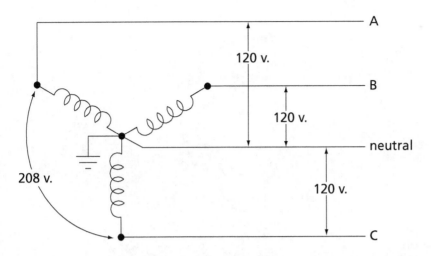

Much of the voltage for the outlets in an industrial plant comes from Wye transformers. The single pole breakers that protect a lot of the outlets are 120 VAC single phase from a Wye transformer secondary.

Illustration 8.3
208 VAC three phase Wye, 120 VAC single phase

Delta

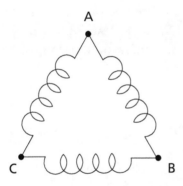

Illustration 8.4
three phase delta

Currents for delta transformer hookups are related to 1.732 (the square root of 3), but the voltages are not. Delta voltages most often have a 50 percent or 200 percent relationship to each other, depending whether you are looking at it from the perspective of the larger or smaller voltage. If you have a three phase delta transformer secondary hookup that is 480 VAC phase to phase and pull a neutral [neutral conductor] off the center point of a phase, you will get 240 VAC.

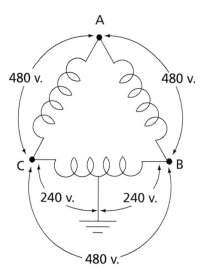

Illustration 8.5
480 VAC three phase delta, 240 VAC single phase

Some of the voltages for 240 volt loads in an industrial plant come from delta transformer hookups. The single pole breakers that protect some of the European equipment loads are 240 VAC single phase. Again, with currents (not voltages), transformer hookups in delta are related to 1.732 (the square root of 3).

If you have a three phase delta transformer secondary hookup that is 240 VAC phase to phase and pull a neutral off the center point, you will get 120 VAC phase to neutral.

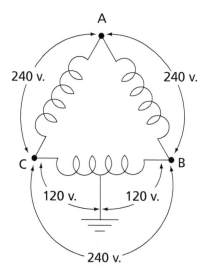

Illustration 8.6
240 VAC three phase delta, 120 VAC single phase

A small minority of the voltages come from this 240 VAC three phase delta secondary for the outlets in an industrial plant. The single pole breakers that protect some of the outlets are 120 VAC single phase from a delta transformer secondary. Please note that most industrial plant 120 VAC outlets come from a Wye secondary because it is a more efficient use of transformer resources.

In delta, you do not want to have a hookup (phase to neutral) that looks like X in illustration 8.7 as this will give you what is called a "bastard voltage." This voltage will not match any equipment that you have. It usually only shows up in troubleshooting situations where somebody has hooked up something incorrectly.

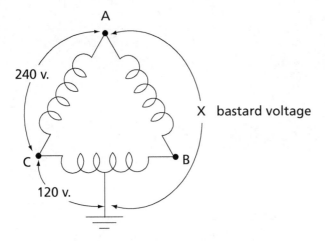

Illustration 8.7
delta "bastard" voltage
(X marks the spot)

Three Phase Motors

When working in industrial plants, the most common type of motor hookup that you will run into is not single phase. The most common motor hookups by far are the three phase Wye and the three phase delta hookup. The main difference between motor hookups and transformer hookups in the three phase world are the number of coils and leads involved. In general, a three phase motor has six coils and twelve leads, and a transformer has three coils and six leads on each primary and on each secondary.

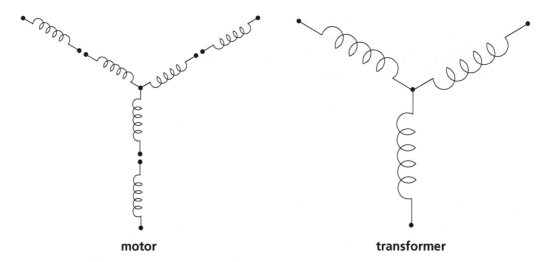

motor   transformer

**Illustration 8.8
Wye motor and Wye
transformer secondary**

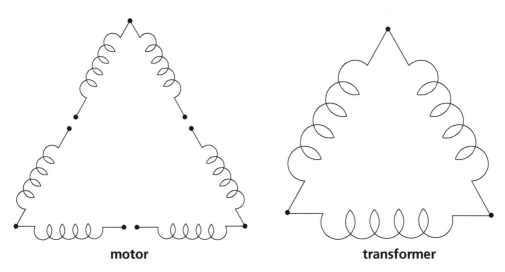

motor   transformer

**Illustration 8.9
delta motor and delta
transformer secondary**

Please note there are often more then six hookups available on three phase transformers but that these are called taps. They are used to increase or decrease the voltage output of the transformer due to a long wire run, etc.

The most common three phase motor hookup in industry is a six coil, nine lead three phase hookup in Wye or delta so that you have the flexibility of running the three phase motor on two different voltages. This is called a high voltage and a low voltage hookup or, sometimes, a series and a parallel hookup.

8 · Wye and Delta

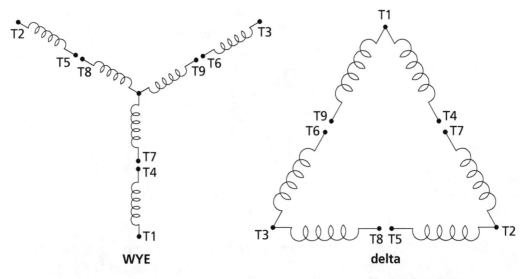

**Illustration 8.10
nine lead Wye and delta three phase motor terminals**

The low voltage hookup for a nine lead Wye motor is T1 to T7, T2 to T8, T3 to T9, and T4 to T5 to T6.

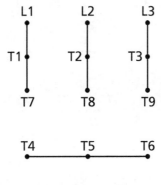

low voltage WYE

**Illustration 8.11
low voltage Wye hookup**

The high voltage hookup for a nine lead Wye motor is T4 to T7, T5 to T8, T6 to T9, and T1, T2, T3.

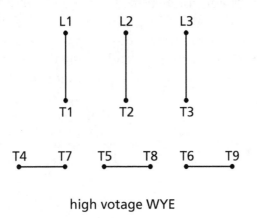

high votage WYE

**Illustration 8.12
high voltage Wye hookup**

8 · Wye and Delta

The low voltage hookup for a nine lead delta motor is T1 to T6 to T7, T2 to T4 to T8, T3 to T5 to T9.

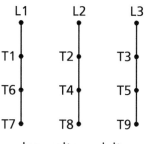

Illustration 8.13 low voltage delta hookup

low voltage delta

The high voltage hookup for a nine lead delta motor is T4 to T7, T5 to T8, T6 to T9, and T1, T2, T3.

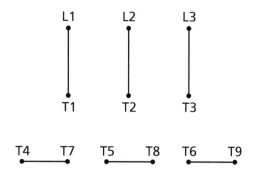

Illustration 8.14 high voltage delta hookup

high votage delta

Think about it this way: the only difference in a three lead, six lead, nine lead, or twelve lead three phase motor is how many of the leads are hooked up inside of the motor. Any of the other leads on the motor not identified as T1–T12, or as L1–L3, are usually heat sensor probe leads or motor brake leads.

Three phase motors are most easily tested by turning the power off and testing the motor at T1, T2, and T3 on the bottom of the motor starter contactor.

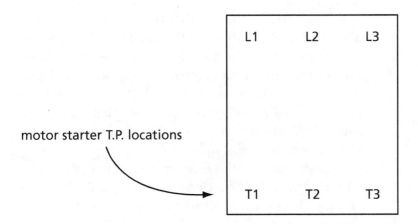

**Illustration 8.15
the typical motor starter
test points (t.p.)**

Test for open, short, leakage, and value (like you did in exercise 5.1) at the bottom of the motor starter. If everything passes, then you know that your motor is OK and that the problem is elsewhere. If one of your tests at the motor starter fails, then you need to do the same test again at the motor. If that second test at the motor passes, then the problem is in the conduit between the motor starter and the motor. If the test at the motor fails, the problem is in the motor.

Single Phase Transformers

In the single phase world, the AC waveform is called a sine wave and has this shape and value:

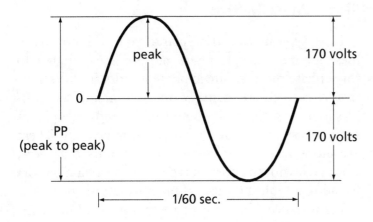

**Illustration 8.16
120 VAC waveform**

As you can see, the wave actually reaches a height of about 170 volts positive and about 170 volts negative. What you need to remember in electrical troubleshooting is that equipment like solenoids and motors live in a time realm that sees 120 volts in this situation. Officially this is called 120 volts rms (root mean square). Electronics live in a time realm that sees a wave that is 340 volts high.

It is sort of like the difference between rust and a stick of dynamite. That difference is time. Both are examples of oxidation, but one happens much faster than the other.

The distance from the zero line to the top of the wave form is called the peak voltage, in this case 170 volts. The rms value (also called the effective value) of the same waveform is 120 volts. The relationship between the two is 1.414 (the square root of 2). A good voltmeter would read 120 VAC. What that means in the world of seconds, minutes, and hours is that the equivalent heating capacity of this 340 volt high AC waveform (called pp or peak to peak) is equivalent in heating capacity to a 120 VDC circuit. In other words, if you hook up equally sized heaters, (one to a 120 VAC rms room outlet and the other to a 120 VDC forklift battery) they will both heat up just as quickly. They would also heat up an equal amount. Hot tea anyone?

Single Phase Motors

In a typical industrial plant, you will not troubleshoot many single phase motors as compared to three phase motors. Single phase motors (not the control circuitry) are actually more difficult to troubleshoot than three phase motors for most people, as single phase motors are not built symmetrically. There are more varieties of single phase motors out there, compared with the normal types of three phase motors found in a typical industrial plant situation.

For instance, if a single phase motor were built symmetrically, it would not turn when you put the juice on. The motor would just sit there and hum. You would have to grab hold of the shaft after the power is on and give it a twist in order to get it moving. It is not a very safe situation.

In order to start a single phase motor without touching the shaft, the designers of these motors have to use some kind of a trick to get it to rotate. On small fans like a 1/16 hp bathroom fan exhaust motor, they use a notch and a piece copper wire to cause a kink in the magnetic field. This kicks the motor shaft to the right or left when the motor starts. This is known as a shaded pole motor.

Most other types of single phase motors use a second coil of wire in the motor called a start coil to cause a kink in the magnetic field and kick the motor shaft to the right or to the left when the motor starts. These are known as split phase, capacitor start, etc. single phase motors. The name of the motor is dependent on the type of hardware that the designer used to get the motor to turn in the right direction.

Typical problems with single phase motors not already covered in 1-800 # troubleshooting or in the control circuitry already done in the exercises so far would be: (1) dust in the contact of the centrifugal switch, (2) welded centrifugal switch contacts, or (3) a bad starting or running capacitor.

K Factor

The K factor of a transformer or other electrical device or wiring hookup is related to how much the electronic equipment and all those associated switching power supplies bends the pure AC sine wave in an industrial setting out of shape. These electronic supplies will cause overheating in electrical wiring, electrical equipment, and transformers because of high harmonic currents. The only way around this problem in most situations is to derate the current levels in electrical equipment, or to supply K rated transformers and/or oversized neutrals in electrical circuits. K transformer matching is done by picking a transformer with a K rating equal to or greater than the K rating of the electronic equipment that you are using in your plant.

Exercise 8.1

Build the following circuit using 120 VAC components if you are following along at work or in a school lab setting. Use 24 VAC components if you are building the circuit at home. In addition to items from previous circuits, you will need a selector switch and a three phase motor. If you are building this circuit at home, use a single phase motor as your load instead of the three phase motor.

Illustration 8.17 run jog motor starter circuit using a motor starter and a selector switch

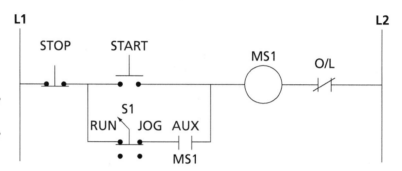

***** *Warning* note: The following exercise requires that your hand be in close proximity to live power.
*****

Exercise 8.2

If you are building this circuit at work or in a school lab, get a clamp-on ammeter. After setting the scale of the clamp-on ammeter properly, measure the actual current flow on any one of the incoming power leads to the motor while the motor is running.

Illustration 8.18
the clamp on ammeter

The good news about the clamp-on ammeter is that you do not have to cut the power leads to the motor and insert an ammeter in series with the current flowing to the motor as was necessary in the past.

Illustration 8.19
a series ammeter

Time Delays 9

Time delays are used extensively in industry in control circuit situations in order to provide maximum flexibility in designing and implementing various circuit combinations. One of these time delay circuits is probably quite familiar to you. When you drive somewhere at night and reach your destination, you shut everything off and exit your car. Some cars have lights that will stay on for perhaps an extra five seconds or so. They are wired this way so that you can see your way to the door of your destination, even in the dark. After a brief period of time, the lights dim and go out. This is an application of a time delay circuit.

On Delay

The most common time delay is the "on delay." This is also known as a delay on make, (DOM), timer on delay, TON, delays coming on, and the symbol:

N.O. on delay   N.C. on delay

Illustration 9.1 on delay schematic symbol

The most widely utilized on delay circuit is probably the burglar alarm circuit. A protected area, such as a house, generally has sensors on the windows and the doors. If a crook breaks into the house and trips a sensor, the alarm goes off after a delay of perhaps twenty seconds. The automatic telephone dialer then calls the police. This time delay is called "on delay." because the alarm delays coming on for twenty seconds or so.

When the homeowner comes back to his protected house and opens the front door, this action trips a sensor. After a delay, the alarm will go off and the automatic telephone dialer will call the police, unless the homeowner does something first. He must go to the security interface pad and turn the system off with a code before the alarm sounds. This time delay is also called "on delay" because the alarm delays activating unless you turn it off first.

For example, in an industrial setting, this circuit may be used to apply hot melt glue to a shipping box flap. The box comes down a conveyor and a sensor detects the front edge of the flap. After a delay of perhaps a ¼ second, the hot melt is shot out onto the flap. This time delay is used to keep the hot melt from dripping off the front edge of the flap and getting on the outside of the box. If you delay the start of the hot melt glue gun by ¼ second, the glue pattern on the flap starts maybe ½ inch back from the edge and goes for several inches. Hence, no leakage is apparent on the product's box, because the time delay starts the flow of glue after the front edge of the flap toggles a sensor.

Off Delay

The next most common type of time delay is the "off delay." This is also known by the names delay on break, (DOB), TOF, timer off delay, delays going off, and the symbol:

N.O. off delay   N.C. off delay

Illustration 9.2 off delay schematic symbol

The most widely utilized on delay circuit is probably the one mentioned at the top of this chapter, the car. When you reach your destination you shut everything off in the car, but some of the lights still stay on for a few seconds anyway. This time delay is called "off delay." The circuit delays turning off some of the car's lights for a number of seconds.

In an industrial setting, this circuit may be used to keep a fan running for several minutes on a cooler for hydraulic oil after the main machine is shut down. This would ensure that the fluid is left at a more reasonable temperature in case the machine needs to be restarted. This time delay is also called "off delay." The circuit delays turning off the cooling fan for several minutes.

There are also two types of off delay in industry. The most common type is the one based on electronic circuitry. This delay is manifest as follows: the output remains OFF when everything is in the rest position. The output immediately turns ON when the input is triggered to be in the active mode. The output remains ON when everything is in the active mode. The output turns OFF five seconds (in this example) after the input is triggered OFF.

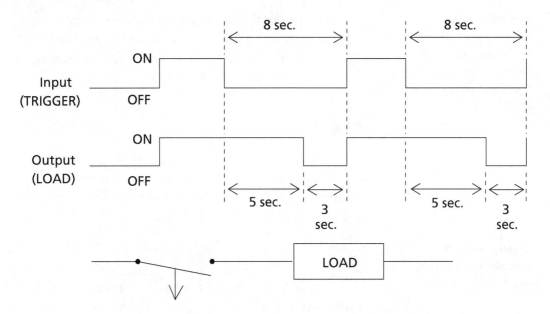

The least common type of off delay is the one based on the on delay timing can. This off delay utilizes a normally closed contact of an on delay timer. This delay is manifest as follows: the output remains ON when everything is in the rest position. The output turns off 5 seconds after the input is triggered ON (in this example).

Illustration 9.3
timing chart of the most common version of off delay (8 second trigger, 5 second delay example)

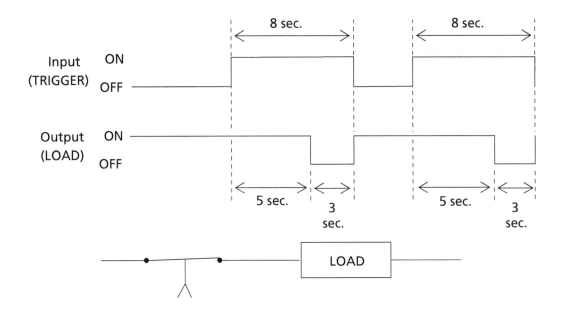

**Illustration 9.4
timing chart of the least common version of off delay (8 second trigger, 5 second delay example)**

This type of off delay that uses a NC on delay contact has not been commonly used since the 1980s, but it still shows up on control prints and in electrical exams in the present day. You just need to be aware of it.

One Shot

The third most common type of time delay is the one shot. This is also known as anti chatter relay, debounce, OS, OSR, one shot rising edge, OSF, one shot falling edge, and the symbol:

ONE SHOT

**Illustration 9.5
one shot schematic symbol**

There is no set independent symbol for a one shot timer symbol on a NEMA print. Usually, it is done by notation on a standard contact symbol.

The most widely utilized one shot circuit is your computer keyboard. One shots are used extensively in man/machine interfaces, as most machines in the computer era "think" much faster than human beings can. In the case of a computer keyboard, if there is no debounce circuit (one shot) in the system and you pressed the letter Q, for example, you would have anarchy.

The letter Q would hit bottom the first time you pushed it and click, then bounce and hit a second time, then bounce and hit a third time, etc. It might bounce a hundred times, and you'd end up with a screen full of Qs in a heartbeat because the computer is working much faster than your finger is moving on the keyboard. Computer programmers implement a one shot circuit (often called de-bounce in the computer world) in either software or hardware so that no matter how many times the letter Q bounces (the keyboard input) within a certain time frame (say 30 milliseconds), it will only register as only one key press (the keyboard output). This type of time delay is called "one shot." he letter Q is passed through the computer to the screen on the first click, but during all the other bounces from the letter Q, the circuit has been effectively disarmed for 30 ms.

Other Time Delays

Almost all other types of time delay circuits are made from the three main types of time delays mentioned above. For example, a repeat cycle timer would be used for a flashing light. This is a combination of an on delay and an off delay timer circuit. The input and the output waveform would look something like this:

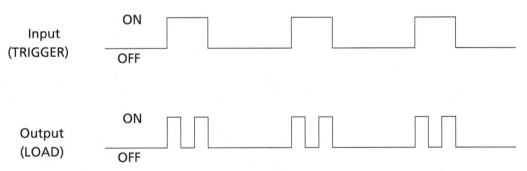

Illustration 9.6 repeat cycle timer waveform example

Timers for industry come in three formats. You can buy a specific timer in an individual unit or "can" with a plug in socket, for perhaps $30 or so. You can buy a multi-timer unit that has many functions in an individual can for about $60. You can buy a PLC as an individual unit for $120–$900. With the PLC, you can have literally hundreds of timers available to you in software and a number of outputs and inputs available to you on the PLC itself.

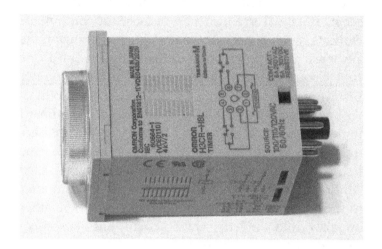

Illustration 9.7 discreet timer

PLC Timers

The most inexpensive way to get a lot of timer functions is to get a small PLC with several inputs and outputs. For example, a hardwired control relay might cost about $15, a timer about $30, a counter about $120, and a sequencer about $200. A $1400 PLC might have in software about ten or twenty sequencers, close to one hundred counters, several hundred timers, and maybe a thousand control relays available on it. You do the math.

Exercise 9.1

Build the following on delay circuit using 120 VAC components if you are following along at work or in a school lab setting. Use 24 VAC components if you are building the circuit at home. In addition to items from previous circuits, you will need an on delay timer can.

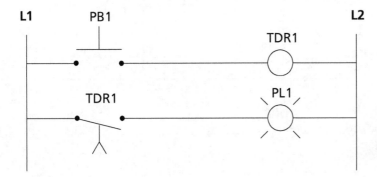

Illustration 9.8
on delay circuit using a pushbutton switch and a light

This circuit uses a timing can that has an active sensor circuit. This means that the sensor is hooked to regular voltage and current.

Exercise 9.2

Build the following off delay circuit using 120 VAC components if you are following along at work or in a school lab setting. Use 24 VAC components if you are building the circuit at home. In addition to items from previous circuits, you will need an off delay timer can with a dry contact for a trigger.

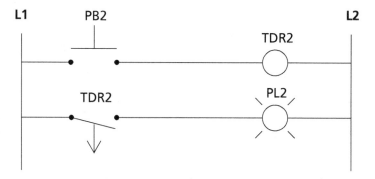

Illustration 9.9
off delay circuit using a pushbutton switch and a light

This circuit uses a timing can that has an [intrinsically safe] sensor circuit called a dry contact trigger. This means that the sensor is hooked to internal low voltage and no current (although actually, the current is on the order of micro amps). This also means that the sensor can be placed in an explosive atmosphere without any worry about fire or explosion from sparking if there were a short in the sensor circuit.

Exercise 9.3

Build the following off delay circuit using 120 VAC components if you are following along at work or in a school lab setting. Use 24 VAC components if you are building the circuit at home. In addition to items from previous circuits, you will need an on delay timer can.

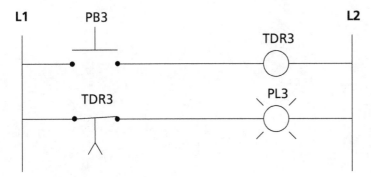

Illustration 9.10
off delay circuit using an on delay timing can, a pushbutton switch, and a light

This circuit uses an on delay timing can that has an active sensor circuit. This means that the sensor is hooked to regular voltage and current.

Three Phase Reversing and Interlocking Starters

10

In many situations in industry, removing a jam or clearing out a machine requires temporarily reversing the machine's direction. One of the ways that this is accomplished is by using a three phase reversing motor starter and interlocking the two contactors on the starter in one of three ways: mechanically, electrically, or using a PLC.

Mechanical Interlocking

The three phase starter generally has two contactors and one overload relay. These two contactors are called F and R, or forward and reverse. They generally are built side by side and have a mechanical interlock installed so that even if electricity is accidentally applied to both coils, the mechanical interlock will only allow

Illustration 10.1
starter mechanical interlock

one of the coils to pull in at a time. It will prevent the other coil from being active at the same time.

In some critical circuits, the forward and reverse pushbutton switches in the circuit are also mechanically interlocked. This is done to keep both switches from being active at the same time, even if they are pressed at the same moment. This creates a backup mechanical safety position. This would show up on a schematic print as a series of dotted lines between the forward and reverse pushbuttons.

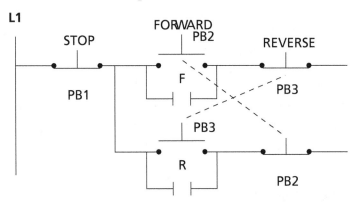

Illustration 10.2 mechanical pushbutton interlock control print elements

Electrical Interlocking

The three phase starter generally has two contactors. These two contactors are called F and R (forward and reverse). They generally are built side by side and have many auxiliary contacts installed (both N.O. and N.C.) so that the auxiliary contacts can be wired up as interlocks and latching elements. This is done so that even if electricity is accidentally applied to both the F and R control circuits at the same time, the electrical interlocks

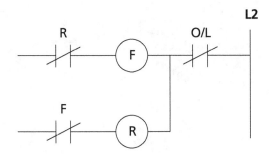

Illustration 10.3 electrical interlock control print elements

10 · Three Phase Reversing and Interlocking Starters

will only allow one of the coils to pull in at a time. It will prevent the other coil from being active at the same time.

PLC Interlocking

Another way of doing the electrical interlocking for three phase motor starters is to use a PLC to interlock the two contactors of the reversing starter in software.

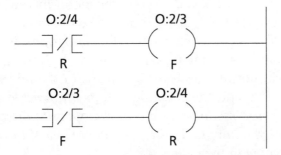

Illustration 10.4
PLC electrical interlock control print

As you can see, the control print in illustration 10.4 looks almost exactly like the control print in illustration 10.3. The main difference on the PLC is how you wire up the circuit (see 10.5). The circuit itself is in the software that you use on the PLC (10.4). The main working difference in the hardwired version (10.3) is that the overload contact is also in the control circuit.

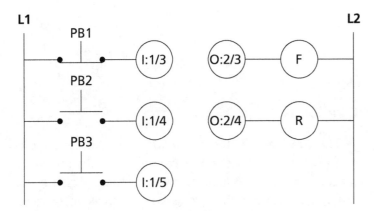

Illustration 10.5
PLC wiring diagram

In real life the actual reversing circuit will use some combination of these interlocks in order to provide a higher degree of safety for the operator and the machine.

Power Circuit

What you are attempting to do with the power circuit of a three phase motor is to have the three leads L1, L2, and L3 go to the motor in the correct order through the F (forward) contactor and have the motor rotate in a CW (clockwise) direction, for example. What you are attempting to do in the power circuit of the three phase motor with the R (reverse) contactor is to have the three leads L1, L2, and L3 go to the motor in a switched lead order (L3, L2, and L1) and have the motor rotate in a CCW (counter clockwise) direction for example.

If you look at the power circuit in illustration 10.6, you will notice that lead L2 remains in the middle of both the F and the R configuration. *In order to get a three phase motor to rotate in the opposite direction, all that you have to do is to switch any two of the three leads going to the motor.*

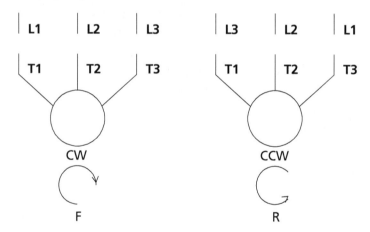

Illustration 10.6 three phase motor reversing hookups

Overloads

The overloads in the overload relay in these circuits and in all three phase motor circuits generally are either mechanical and thermal, or are solid state. The biggest difference between the two is convenience or aggravation. It all depends on the policies at your workplace. Overloads are designed to protect motors within an electrical timeframe of seconds and minutes from burning out because of excessive heat, which could be caused by too much ambient heat, or a jammed up product, etc.

Fixed Mechanical and Thermal Overload Protection

This type of overload protection (which is typically NEMA) for your three phase motors offers the best type of result in a work environment where a lot of people would have access to the cabinets where the motor starters are located. Someone would have to go through a great deal of effort to replace the heaters (overloads) in a unit if they want to "bump up" the trip point of the overload relay. They would have to find higher rated heaters for that particular manufacturer, go to the storeroom to get them, shut off the machine to install them ... you get the picture. This tends to bring more focused troubleshooting to bear on the problem. You cannot get rid of motor trips just by turning up the trip point of the overload relay. It takes more effort to increase a trip point. You might actually have to investigate the real cause of a motor trip.

Adjustable Mechanical and Thermal Overload Protection

This type of overload protection (which is typically IEC) for your three phase motors offers the best type of result in a work environment where a just a few people would have access to the cabinets where the motor starters are located. The restricted access keeps the operators from increasing the overload relay trip point too easily. If they want to "bump up" the trip point they need to go through a formal procedure of some sort with the electricians or the maintenance folks in the plant.

The negative side of this situation is that if you have a lot of people that have access to the motor starter cabinets, then you have a very good chance that someone will change the trip point of the overload relay. This can result in burned out motors and troubleshooting aggravation, as an incorrect trip point can mask a problem like a bad bearing.

Adjustable Solid State Overload Protection

This type of overload protection (which is typically IEC and NEMA) for your three phase motors offers the best type of result in a work environment where a just a few people would have access to the cabinets where the motor starters are located. If someone wants to "bump up" the trip point they need to go through a formal procedure of some sort with the electricians or with management in the plant. The same negatives apply here as in the adjustable mechanical and thermal units enumerated above.

Exercise 10.1

Build the following forward and reversing **control circuit** using 120 VAC components if you are following along at work or in a school lab setting. Use 24 VAC components if you are building the circuit at home. In addition to items from previous circuits, you will need a three phase reversing starter and additional pushbutton switches.

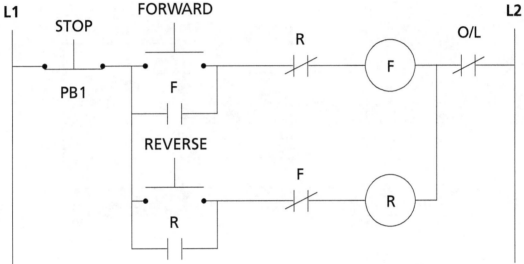

Illustration 10.7 forward and reverse circuit using a three phase motor starter

Exercise 10.2

Build the following forward and reversing **power circuit** using 240 VAC or a 480 VAC three phase components if you are following along at work or in a school lab setting. If you are trying to build this circuit at home do Exercise #10.3 instead. In addition to items from previous circuits, you will need a working three phase motor.

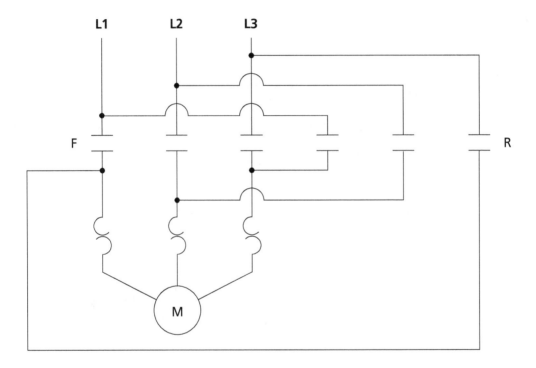

**Illustration 10.8
forward and reverse
power circuit using a
three phase motor**

Exercise 10.3

Build the following forward and reversing **power circuit** using 120 VAC single phase components if you are following along at home, work, or school. In addition to items from previous circuits, you will need a single phase motor with a centrifugal switch.

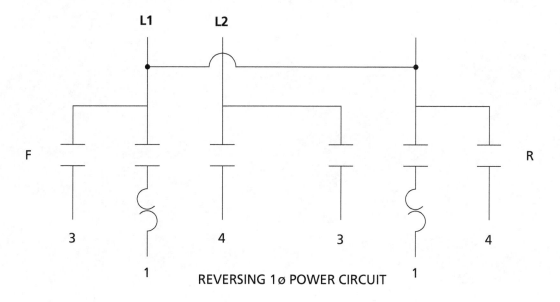

REVERSING 1ø POWER CIRCUIT

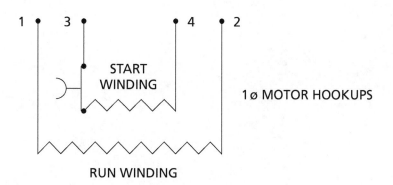

1ø MOTOR HOOKUPS

To reverse a single phase motor with a centrifugal switch, simply reverse the leads of the start winding (the one with the switch on it). Leave the run winding alone.

Illustration 10.9 forward and reverse power circuit using a single phase motor

Variable Frequency Drives (VFDs)

11

In industry, if you need to adjust the speed of a three phase induction motor, reverse the direction of a three phase motor, ramp the speed up and/or down on a three phase motor, or do any number of other things by changing the parameters of a three phase motor, then the VFD is for you. Variable frequency drives (VFDs) are very common in industry because of three main reasons. Firstly, they are very flexible when it comes to speed control (variable, manual, remote, computer controlled, etc.). Secondly, they work with standard three phase AC motors. Lastly, they are extremely energy efficient if you need to run steady speed and variable torque items like blowers, pumps, and fans at less than their maximum speed (for example, at 75 percent of full speed).

VFD

The typical VFD takes incoming AC energy (typically three phase in an industrial plant) and converts it all to DC energy. It then converts all the DC energy back to AC three phase energy. On the surface, this may seem like a waste, but there is a method to this madness. This method gives the machine operator electronic control (along with the option of computer and PLC control) of various aspects of the outgoing AC waveform of the VFD. That waveform then goes into the input terminals of the machine's three phase motor.

The VFD generally has manual controls on the unit and it also has plug-in ports for computer and/or PLC control cables. Let's take a look at a typical VFD and the flexibility it gives you.

134 11 · **Variable Frequency Drives (VFDs)**

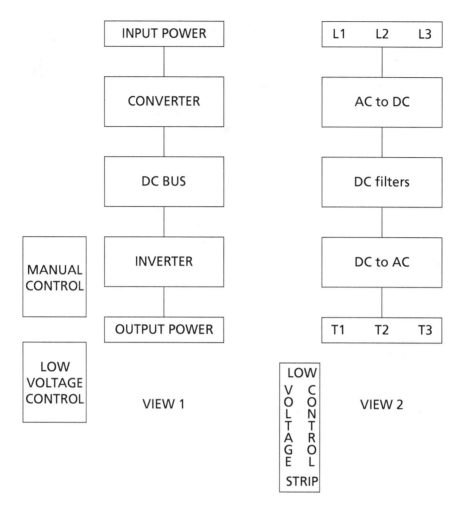

**Illustration 11.1
VFD block diagram
(two views)**

L1, L2, and L3

This is where the power is hooked up to the top of a typical industrial VFD. Most VFDs do not care if you hook up either three phase power (typically 480–600 VAC and below) to the top of the unit at L1, L2, and L3 or hook up single phase power (typically 208 VAC and above) to the top of the unit at L1 and L2. *The VFD needs enough AC energy at the input to run your motor at the output. As long as the energy is within the parameters of the electronic components used to build the VFD (like voltage levels), then the input can have three phase or single phase power.* What this means in practical terms is that if you have 230 VAC three phase wiring in a section of an industrial plant and you want to install a ¾ horsepower (H.P.) 460 VAC three phase blower motor, then you are fine. All that you need to buy is an inexpensive $200 VFD and install it. You do not need to run an expensive 460 VAC circuit into that area of the plant for the blower motor.

What this means in the commercial area is something like this: a farmer, for example, wants to pump water in an irrigation system and has the option of upgrading to a 2.5 H.P. pump that uses 550 VAC 50/60 Hz. three phase from a surplus European system. All the farmer has available is 240 VAC single phase on his farm. No problem. He can get a VFD unit in order to run that pump with 240 VAC single phase on the input of the VFD and 550 VAC three phase on the output of the VFD.

Flexibility is the key. The VFD just needs AC energy on the input. It converts all the incoming AC energy to DC energy. It then converts it back to AC energy in order to run your motor.

Converter (AC to DC section)

Within the converter of the VFD, all the incoming AC energy is converted to DC energy. The voltage and the current have a lot of ripple in it at this point, which can be a problem.

DC Filter (DC bus section)

Within the DC filter of the VFD, all the DC energy is run through a bank of capacitors to remove all of the ripple in the waveform. This area of the VFD is extremely hazardous to troubleshoot, even with the power off. The bleed-off circuit may have failed and the capacitors could retain a lethal amount of charge.

Inverter (DC to AC section)

Within the inverter of the VFD, all the steady (straight line) DC power is electronically manipulated to give the output a three phase waveform that was set via the parameters in software or in hardware on the VFD. This is the area of the VFD where the constant ratio of volts to hertz (V/Hz) is maintained for the motor that you are using along with all the other parameters that you want to set for your particular industrial situation.

T1, T2, and T3

This is where the power is available for hookup to the three phase motor that you will be running.

Low Voltage Control Strip

The low voltage control strip is the area of the VFD where the low voltage inputs and outputs exist for 4–20 ma interface, 0–5 VDC interface, dry relay contacts, etc. These are some of the voltages and currents that may be available to use on a particular VFD. In a lot of control situations in the industrial and commercial area, low voltages and low currents are used for safety reasons.

24 VAC Control (HVAC)

Within the HVAC area, the thermostats in a typical office or home are 24 VAC. This is done so that if your kids are playing "catch" in the house with a foam ball and someone collides with the thermostat on the wall and breaks it, then nobody will be electrocuted. It would be easy to run 240 VAC through that thermostat directly from your air conditioner's compressor case and back to the hallway of your house for your control circuit, but it is safer to use a transformer and run the thermostat on 24 VAC instead. If little Johnny gets a jolt, then everybody is going to be the loser. After all, who would want 240 volts lurking just below the surface of the pushbutton on your front doorbell, when 16–24 VAC works just fine?

0–10 Volt DC Analog Control

This type of analog control is used in circuits as diverse as dimmer circuits for lights and external potentiometer control of pump circuits through a VFD.

0–5 Volt DC Analog Control

This type of analog control is used in computer control circuits and PLCs, usually in the form of A to D (analog to digital) and D to A (digital to analog) circuits.

5 Volt DC Digital Control

This type of electronic digital control is used in computer control circuits, usually in the form of on-off circuits.

4–20 ma. DC Analog Control

This type of analog control is used in process control circuits. It is usually transmitting variable information on pressure, temperature, weight, etc.

4–20 ma. DC Digital Control

This type of digital control is used in computer circuits usually in the form of an on-off signal. This type of control circuit was originally used with teletype machines.

Installation Issues (Small Terminals)

Generally speaking, VFDs are installed by electricians but designed by electronic engineers. What this means on the installation side of things is that the terminals on the VFD are going to be too close together for the way that a typical electrician operates. They are also going to be too close for installation for the typical electronic technician in an industrial three phase environment, for that matter. Problems including a strand of wire (a whisker) from a three phase line shorting out on another terminal or a wire hooked to the wrong terminal due to teeny, tiny, little labels are very real.

Installation Issues (Confusing Lingo)

The typical information booklet that comes with a VFD unit is oriented to computer people and electronic techs. The typical electrician may be hard pressed to understand all the choices that are available for setting and resetting parameters on the VFD unit until he gets used to working with them. The good news is that on average, you will need only about ten of the six hundred or so commands and sub commands that are available for most VFD motor controllers. This command subset usually consists of reset, nameplate volts, nameplate amps, nameplate frequency, nameplate motor rpm, acceleration time, deceleration time, run rpm, source of control, and reverse.

Installation Issues (Whiskers)

Often on VFDs the terminals are too close at the L1, L2, and L3 area and at the T1, T2, and T3 area. It would be very prudent to take a few extra minutes to make sure that all of the strands of wires on these terminals are under their designated screw terminal. It is very easy to have a wayward whisker of a stranded wire stick outside its designated terminal area and touch the next area over. If the VFD is then turned on, you will more than likely blow the unit due to a dead short.

Installation Issues (Wrong Voltage)

On a VFD, the input and output terminals need to be connected to the right voltage and the circuit needs to be able to supply all the current needed. Check the label on the unit. A lot of different VFD models from the same manufacturer look exactly the same on the outside. The only difference in some of the VFD controllers are the components on the inside of the units. The outside of the unit looks the same in many situations, so check the label.

Almost all VFDs have a low voltage terminal strip for 4–20 ma sensors or for 0–5 VDC interfaces, dry contacts, etc. You will fry part of your unit if you mistakenly hook a 110 VAC wire one terminal over to where the 5 VDC input exists on the VFD. You can not afford to hook a wire from a motor control circuit intended for a N.O. dry contact on the VFD and instead hook that 110 VAC line to a 5 VDC terminal.

Installation Issues (Wrong Parameters)

While the guide for the VFD unit can be confusing, there is one thing that can really throw you off. One of the ten or eleven major parameters could be entered wrong (nameplate volts, nameplate amps, nameplate frequency, nameplate motor rpm, acceleration time, deceleration time, run rpm, source of control, reverse, or trip point). Hey, it happens. Do this parameter setup:

A. Hook single phase power to L1 and L2 on your VFD. Hook the smallest three phase motor that you can find to T1, T2, and T3 on your VFD.
B. Turn on the power to the VFD. In the software, find the reset code to take the unit back to its factory settings. Punch it in and reset the unit. Then, set the eleven parameters listed above for the small three phase motor that you hooked up.
C. Run the motor using the stop start switch on the VFD unit and using the variable speed pot. (the potentiometer or the variable resistor) on the VFD unit.

You need to be able to do this setup successfully or to troubleshoot this setup successfully before you do the exercises in this chapter.

Installation Issues (Bad Plant Input Power Circuit Voltage Levels)

In some industrial plant situations, you have considerable sag in the plant voltage, along with brownouts, drop outs, or voltage quality issues. You need to make sure that you choose a more robust VFD and/or use line conditioners (like chokes) for this type of situation. You need to check the specs of the VFD to see if it can meet most of these challenges, if they exist in your plant.

Exercise 11.1

Build the following start stop **control circuit** using 120 VAC components, if you are following along at work or in a school lab setting. Use 24 VAC components if you are building the circuit at home. In addition to items from previous circuits, you will need a VFD unit.

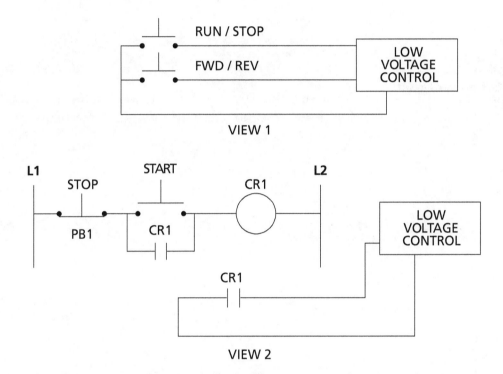

Build the following **power circuit** using a three phase 208–240 VAC motor and a VFD that has a three phase 240 volt output. Use a VFD that can use 120 VAC at the input if you are following along at home. Use a VFD that can use either 208 VAC single or three phase at the input if you are doing this circuit at work or in a school lab setting. In addition to items from previous circuits, you will need a three phase 208–240 VAC motor.

Illustration 11.2 stop start control circuit for a VFD

Exercise 11.1

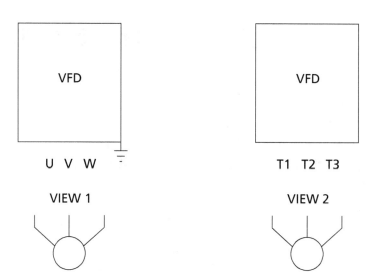

Illustration 11.3 stop start power circuit for a VFD

Make sure that you pay close attention to the installation issues mentioned previously in this chapter. Take your time.

Exercise 11.2

Build the following start stop **control circuit** using 120 VAC components with external switches and potentiometers, if you are following along at work or in a school lab setting. Use 24 VAC components if you are building the circuit at home. In addition to items from previous circuits, you will need a pot (potentiometer).

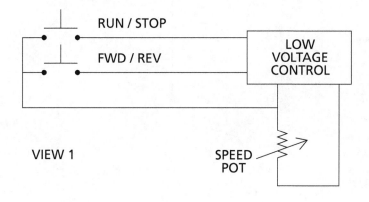

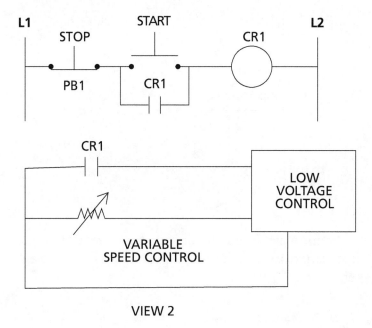

Illustration 11.4
stop start control circuit for a VFD with external switches and an external potentiometer

Exercise 11.2

Build the following **power circuit** using a three phase 208–240 VAC motor and a VFD that has a three phase 240 volt output. Use a VFD that can use 120 VAC at the input if you are following along at home. Use a VFD that can use either 208 VAC single or three phase at the input if you are doing this circuit at work or in a school lab setting.

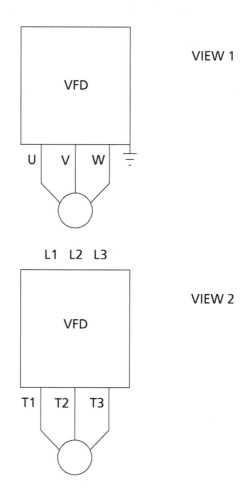

Illustration 11.5
stop start power circuit for a VFD with external switches and an external potentiometer

Again, make sure that you pay close attention to the installation issues mentioned previously in this chapter.

Servos and Intermediate Troubleshooting   12

Machine operations in a manufacturing environment in the present day are more likely than not to have machines that are run by servos. Let's examine what in general is meant by the term, "this machine is run by a servo motor."

Resolvers and Synchros

All servo motors have some type of feedback mechanism. Some motors use resolvers. These are units that resemble a small motor attached to the main shaft of the machine rotor that send feedback (positional) information. They can be thought of as a type of rotary transformer that measures angular rotation in degrees. They tend to have four leads (two or four coils). The three coil (three lead) types are generally called synchros. Of course, there are different sub types, but in general if the feedback is in some multiple or in some sub-multiple of 90 degrees, it is generally called a resolver. If the feedback is in some multiple or in some sub-multiple of 120 degrees, it is generally called a synchro.

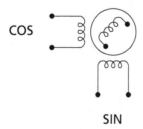

Illustration 12.1 typical resolver transducer unit circuit

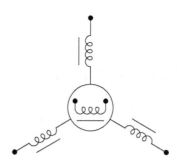

**Illustration 12.2
typical synchro
transducer unit circuit**

Encoders

All servo motors have some type of feedback mechanism, but the most common by far is the encoder. The most recognized encoder is the type where a LED light source looks across a gap at an electronic receiver (like a photo transistor, etc.). In the gap between these two electronic components rotates a plastic disk with alternating dark and transparent sections. By reading, keeping track of, and calculating the alternating transparent and dark bands, the encoder can be used to give the machine in question the absolute or relative position of the motor shaft that is physically hooked to the encoder.

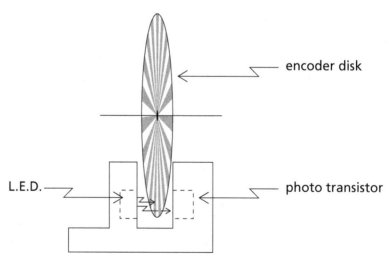

**Illustration 12.3
typical encoder setup**

Stepper Motors

Stepper motors tend to be used in situations where you need to move a light, fixed static load at low acceleration to a repeatable position. A case in point would be the print head position for the letters on an 8½ inch by 11 inch piece of paper in an ink jet printer. The scan head position for a copying machine or a scanner unit would also be an example of use of stepper motors. This does not mean that stepper motors are slow. They can go very fast with low accelerations and decelerations. They can stop or start very quickly with light loads, without overshooting or undershooting their intended position.

Servo Motors

Servo motors tend to be used in situations where you have feedback with encoders and (1) you need to move heavy loads to a repeatable position or (2) you need to move light loads at high accelerations to a repeatable position. A packaging machine would be a good example of the first situation, where you need to move heavier loads to the same position over and over again. A pick and place type of robot that stuffs circuit boards with components would be an example of the second situation where you need to move light loads at high accelerations and decelerations to a repeatable position. A CNC milling machine would be a good example of a combination of (1) and (2). It is a very fast moving, highly accurate machine when the cutting head is moving into position to make the next pass, but can become a slow moving, highly accurate machine when heavy resistance is met in the actual cut.

Intermediate Troubleshooting

When things go wrong, that old expression seems to kick in: "20 percent of your problems will end up consuming 80 percent of your troubleshooting time." In a way that is a good ratio, because it shows that the person doing the work is falling back on some type of troubleshooting system. This allows a better focus on the amount of time on task and ultimately getting the job done well.

We see a lot of situations where something that should take twenty minutes or less to fix can end up taking the better part of a shift. Why? It's usually because the machine operator, the maintenance guy, or the electrician is trying to do what he thought he saw Joe do in a similar situation on that machine in the past. But guess what? Joe skipped over a lot of easy fixes. The first thing Joe wanted to do was to get the voltmeter out and start measuring, without testing the meter on a known source first. He might even get the electrical schematic print out and try to become the big shot.

What you need to do in a manufacturing production troubleshooting situation is to quickly and professionally gather information. Get it from the operators and any mechanics or electricians that have been looking at the problem before you walked up. Act on their information, but do not waste too much time on it. You need to jump to some sort of a troubleshooting system quickly in order to find the problem and fix it in a relatively short time frame.

You have to take everything that anyone tells you with a grain of salt. It is just human nature to make statements such as, "I hooked it up just like the other one," when they did not check the voltage level on the replacement coil because it looked just like the other one. It is just human nature to make statements such as, "I already checked all the fuses and circuit breakers," when they forgot about the three fuses behind the hinge of the machine door.

We use what we call the 1-800 # troubleshooting system as our approach. It doesn't matter what system you use to troubleshoot electrical problems, so long as you have some sort of troubleshooting system that will eliminate and fix the most common electrical problems fairly quickly. This keeps you from spinning your wheels and wasting a lot of time and resources. It gets you out of the business of jumping all over the place

12 · Servos and Intermediate Troubleshooting

trying to follow leads that never seem to quite pan out. We also use the 1-800# system because, if you work in an environment where you can not work directly on live electricity (using voltmeters and clamp on ammeters for example), then you can still fix about 2/3 of all the electrical problems that you will ever run into. This can happen by just using your brain, a system, and an ohm meter.

Intermediate troubleshooting in most electrical situations usually kicks in under four scenarios. (1) When you reach the 70–80 percent level under the 1-800# troubleshooting system, (2) when troubleshooting VFDs, (3) when troubleshooting servos, and (4) when troubleshooting something that you have never ever seen or heard about before.

Using Voltmeters

When using a voltmeter, the biggest mistake that is made, "bar none" is picking up the meter and starting to use it just as it is.

This would be how you would want to use a voltmeter in a 240 VAC control circuit for example:

1. If you are going to use a voltmeter, make sure that it is on.
2. Make sure that the voltmeter is measuring the right thing (in this case AC volts). A lot of voltmeters are actually multi-meters. That means that they measure things other than volts.
3. Make sure that the voltmeter is on the right scale (in this case 500 volts, or automatic).
4. Check the meter on a known source, like an electrical outlet, to make sure that the meter works properly on this scale setting, before you use it on your test item or your test point.
5. Make sure your fingers are behind the nubs or ribs on the two probes when making measurements.
6. Do not wear jewelry, watches, rings, etc. if you are measuring live voltages.

Using Ammeters

Ammeters generally come in two flavors. The original type of ammeter is inserted in a circuit in series. This can be a bit of a problem for someone wanting to measure the current in the individual phases of a three phase motor. You would first shut off the power, then take a "hatchet" or a cutter and sever the L1 line going into the motor, then hook up the ammeter, then turn on the power, take your reading, and repeat everything for L2 and L3. This is very messy and would leave a lot of lines to splice in a box. or wires to run again in a conduit.

Illustration 12.4 traditional ammeter on a three phase motor

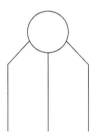
BEFORE

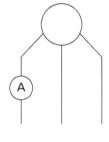
AFTER

A much more elegant way to measure amperage in a wire is to use a clamp-on ammeter.

1. Do not wear jewelry, watches, rings, etc. if you are measuring live voltages.
2. Make sure that the ammeter is on the right scale (and in this case that the needle is free to move or the display is not in lock mode).
3. Put on appropriate PPE for the motor starter electrical box.
4. Open up the lid.
5. Take your readings at T1, T2, and T3 at the bottom of the motor starter.

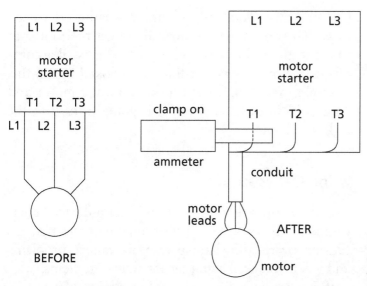

Illustration 12.5 clamp on ammeter on a three phase motor

Using Meggers

A megger is actually an ohm meter that uses a much higher voltage than the 1½ volt or the 9 volt battery that typical comes in the normal analog or digital ohm meter. Meggers use 500 to 1000 volts at low current levels (often done with a hand crank) to stress the insulation in a coil, motor, or conduit run. This is done in order to find when the insulation is beginning to break down. This instrument can catch things before you have a catastrophic failure. This is usually done by testing critical components on some sort of schedule and comparing readings over time. This testing will also confirm where some of your intermittent problems are coming from.

Using Rotation Meters

Rotation meters are used on larger three phase motors that will be hooked to loads that cannot run backwards, even for a very short period of time. The meter is hooked up to the L1, L2, and L3 leads of the three phase motor before it is hooked up to a power source. The shaft is then turned gently by hand. The meter will indicate if the motor is hooked up to rotate

CW or CCW. This will determine whether you need to switch two of the motor leads to get the motor to rotate in the right direction or not. This is all accomplished before any power lines are hooked up. In this example, you do not need to hook up the motor and "bump it" in order to see if it is going to turn in the right direction.

Using Oscilloscopes

Oscilloscopes tend to be used in an industrial environment when you have power quality issues that are causing overheating and premature failure problems and when you are having sensor signal problems.

If you are unfamiliar with the operation of an oscilloscope, you can still see the power quality in a particular circuit by using two voltmeters. You measure the voltage in the circuit with an average voltmeter (the most common type) and then measure the voltage in the circuit again with a true rms meter. If you have no power quality issues, the readings will be almost exactly the same. If you have a bad situation the readings may be off from one another by 30 to 40% or so. Compare the two readings with a known circuit situation or with a table of values. This can also be done with an average and a true rms clamp on ammeter.

The oscilloscope is the best instrument when you need to take a look at the actual waveform. Rugged solid state versions of this instrument are available for field use.

In Conclusion

Your journey through the text to this point will get you headed in the right direction through the main body of industrial motor control information. You will be prepared to speak with and understand the electricians and the electronic techs that you work with if you are a supervisor or a junior engineer. You will be able to upgrade in your current job or to seek a new one if you are a machine operator or a fixer. You will also become a great electrical troubleshooter if you are a maintenance mechanic, an electrician, or an electronic tech, and as a side bonus it pays well.

Appendix

Selected Component Pictures—2 Views

Illustration A.1
HVAC contactor,
24 VAC coil

Illustration A.2
manual motor starter
(Can you spot the
defect?)

156 **Appendix**

Illustration A.3
1/2 horsepower VFD,
120 VAC input, three
phase output

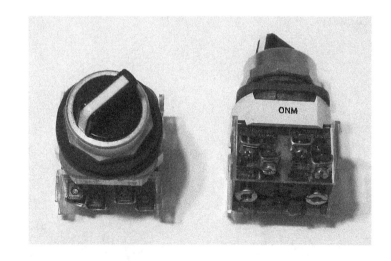

Illustration A.4
selector switch (used
in run jog circuit)

Appendix

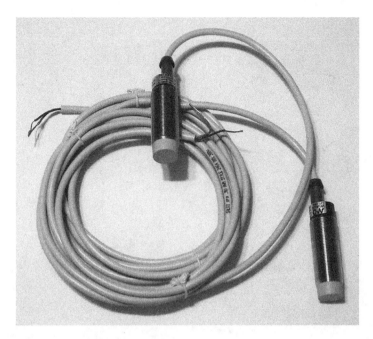

Illustration A.5
two wire AC prox

Illustration A.6
120 VAC control relay

Appendix

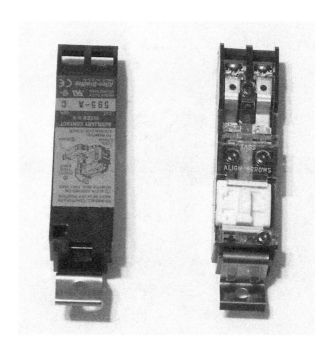

Illustration A.7 auxiliary contact for a motor starter

Illustration A.8 thermal overloads

CPSIA information can be obtained
at www.ICGtesting.com
Printed in the USA
LVHW101827280120
645066LV00012B/478